# A Study of Droplet Deformation

*Hannah Fry*

A dissertation submitted in partial fulfillment

of the requirements for the degree of

**Doctor of Philosophy**

of

**University College London**.

Department of Mathematics

University College London

January 2011

# Abstract

In both engineering and medical applications it is often useful to use the knowledge of the conditions under which adhering liquid droplets appear, deform and interact with surrounding fluids, in order to either remove or create them. Examples include the de-wetting of aircraft surfaces and the process of injecting glue into the bloodstream in the treatment of aneurysms. The particular types of models discussed here theoretically are based on droplets with a large density compared to that of the surrounding fluid. Using this ratio as a small parameter, the Navier-Stokes equations may be simplified, and in view of the nature of the interfacial boundary conditions the droplet may be considered as solid to leading order at any given time step for a certain time scale.

In the first part of the thesis, we study an example of an initially semicircular droplet adhering to a wall for low-to-medium Reynolds numbers (along with simpler test problems). We numerically determine unsteady solutions in both the surrounding fluid and the droplet, coupling them together to obtain a model of the droplet deformation. Analysis within the droplet leads to the identification of two temporal stages, and the effect on large-time velocities is discussed.

The second part of the thesis sees a similar approach applied to a surface mounted droplet completely contained within the boundary layer of an external fluid for high Reynolds numbers. The two-fluid interface for such a regime is analysed using a lubrication approximation within the viscous sublayer of a triple deck structure. Finally, the lubrication is abandoned and we present a fully non-linear solution in air over any obstacle shape, as well as a two-way interacting model of droplet deformation, capable of simulating the free surface of the droplet as it becomes severely distorted.

# Acknowledgements

First and foremost I would like to express my heartfelt thanks to my supervisor Frank Smith, with whom it has been a pleasure to work. I am also sincerely grateful to Sergei Timoshin for his continued interest and guidance, my family and friends for their support and encouragement and to my colleagues within the Mathematics department of UCL, in particular Alex White, Minette D'Lima, Nenna Campbell-Platt and James Burnett.

# Contents

# List of Figures

# Chapter 1

# Introduction

## 1.1    Background and motivation

Thin liquid films, droplets and bubbles are encountered in a wide range of physical conditions. Simple examples include raindrops on car windscreens or air bubbles within a carbonated drink, but there are many examples of medical and industrial application where an understanding of how two fluids interact with one another and their surroundings is important to recreating and manipulating droplet and liquid behaviour.

In the aviation industry, any ice amassing on the fuselage, wings and tail of an aeroplane creates a potential hazard. Unwanted ice can significantly alter the aerodynamics of the flight surfaces: decreasing lift while increasing drag and elevating the risk of the aerofoil stalling. Large pieces of ice breaking off in flight may be ingested by the engine or cause damage to propellers. As such, effective ice removal is vital to aircraft safety. If ignored, catastrophic consequences for aircraft performance may occur, as has been demonstrated throughout history, with 9.5% of all fatal air carrier accidents reporting icing as a contributing factor.

Aircraft are prone to the accumulation of ice both on the ground and in the air. On the ground in freezing conditions, ice or snow can build up, leaving the aerodynamic surfaces as rough and uneven. In flight, as an aircraft passes through the wet base of the cloud, a thin liquid layer of water will form on the wings. If the volume of water passing over the wing exceeds what is blown off by the shear forces, this layer can freeze to form clear ice. As the aircraft continues to climb, the increased altitude and lower temperatures cause the water droplets within the cloud to become super cooled, maintaining their liquid structure, but they will instantaneously freeze on impact with the aircraft, forming solid ice crystals which will adhere to the wing. In such a case rime ice is formed alongside clear ice on the surface of the aerofoil.

Aircraft on the ground may be sprayed with anti-icing fluid to prevent the build up of any ice or snow. After take off the shear forces of the air flow over the wings cause the anti-icing fluid to thin and run off, leaving a clean and smooth aerodynamic surface. An understanding of how liquid droplets and films adhere to surfaces and deform in the presence of a shear flow is important for the design of such anti-icing fluids and the hydrophobic aerofoil surfaces which assist effective de-icing.

Within the brain, 13% of all strokes are caused by aneurysms and arteriovenous malformations (AVMs). Aneurysms involve a localised weakening of an arterial wall, balloon shaped and filled with blood. As the size of an aneurysm increases so too does the chance of bursting, causing severe and potentially fatal haemorrhaging. To avoid the invasive procedure of surgical clipping, recent advancements in the treatment of aneurysms see neurologists feed radiographically guided catheters through an artery in the leg up to the brain. The aneurysm is then filled with a special glue, liquid Onyx, which stabilises the vessel wall as it solidifies, preventing rupture. AVMs, which are an abnormal connection between the arteries and veins within the brain, may be treated in a similar manner, where the glue serves the purpose of cutting off the blood supply to the affected area in a process known as embolization. A knowledge of how two fluids of different densities interact with one another becomes vital to the design of the glue and the success of such a procedure.

In agricultural crop spraying, the objective is to cover all surfaces of the plant evenly with pesticide. Using large amounts of chemical can cause droplets to coalesce and run off from the plant, reducing effectiveness and increasing costs while having a negative impact on the environment. Similarly in industrial spray painting, the aim is to obtain an even coverage using a minimum amount of paint. Creating the desired droplet size to achieve this requires an understanding of how droplets split and merge when passing through a surrounding fluid.

With these and many other motivations in mind, a vast amount of literature on modelling the flows of two fluids has been produced in recent years across many different disciplines, presenting both theoretical and experimental studies. Thomas, Cassoni and MacArthur [71], Politovice [42] and Myers, Charpin and Thompson [40] present work on aircraft icing, White [73] on embolization and Sivakumar and Tropeaon [53] on spray impact. In a more general context, overviews of analytic and computational approaches may be seen in Joseph and Renardy [27] and [28] and Ishii [25], while experimental papers include Dussan V. and Davis [16] Villermaux and Bossa [72] and Bentley and Leal [5].

The free surface or interface between two fluids is of particular interest within the literature as it plays an important role in the dynamics of many two fluid systems. Modelling the evolution of this interface, equivalent in some cases to modelling droplet deformation, is the main aim of this thesis. This task has been tackled computationally by many authors in the past using a wide range of techniques, a selection of which are as follows.

The volume of fluid method (VOF) presented first by Hirt and Nichols [24] and employed, for example, by Renardy, Renardy and Li [48], Renardy, Renardy and Cristini [49], approximates the characteristic functions of the two fluids by using a scalar fraction function. The result may be interpolated to achieve a smooth description of the interface, while conserving the mass of the traced fluid.

A similar approach is that of the level set method, an overview of which may be seen in Osher and Fedkiw [41], which uses an auxiliary function to represent the interface as a closed curve. The numerical simulations employ techniques such as Chorin's projection method as in Croce, Griebel and Schweitzer [12], the Godunov method, Sussman et al. [68] or standard finite difference schemes, Spelt [64].

Other groups such as Harlow and Welch [21] and Bierbrauer and Philips [7] use massless Lagrangian marker particles within the two fluids, where tracking their distribution over time allows the moving interface to be reconstructed.

While the majority of methods treat the interface as sharp, distinct and infinitesimally thin, in reality it is a slowly varying region, a few molecules thick. Drawing from these physical properties, diffuse-interface models such as those outlined in Anderson, McFadden and Wheeler [3] apply a region of continuous variations, with thickness related to a small parameter.

The focus in this thesis is more on the theoretical front, concerning a combination of analysis and computation. In particular, returning to a more standard representation of the interface between the two fluid domains as an infinitesimally thin boundary, recent years have seen a body of theoretical work which capitalises on an assumption of a small density ratio between the two fluids. With this assumption, the leading order velocity term in the less-dense fluid will be zero along the boundary, with the remarkable outcome that, to a first approximation, the interface may be treated as solid at any given time step.

This approach was first proposed by Smith, Li and Wu [59] and then more explicitly by Smith and Purvis [62]. It has since been applied by Smith, Ovenden and Purvis [61] and Hicks and Purvis [22], and is the method upon which the majority of this thesis is based. It is relevant to any two fluid system in which the density ratio is small: liquid films, free surface or attached droplets and for any Reynolds numbers. For ease, we shall refer to our two fluids as air (A) and water (W), where the ratio would be around $1/828$.

## 1.2 The small density ratios approach

In a two dimensional system where the effects of compressibility are ignored, the fluid flows are governed by the conservation of mass and the conservation of momentum. Taking $\mathbf{u}_A^*$ and $\mathbf{u}_W^*$ as the dimensional velocities of the flow within air and water respectively, the first of these is known as the continuity equation:

$$\nabla \cdot \mathbf{u}_A^* = 0, \tag{1.2.1}$$

$$\nabla \cdot \mathbf{u}_W^* = 0, \tag{1.2.2}$$

and the second, the momentum equation:

$$\frac{\partial \mathbf{u}_A^*}{\partial t^*} + (\mathbf{u}_A^* \cdot \nabla)\,\mathbf{u}_A^* = -\frac{1}{\rho_A}\nabla p_A^* + \nu_A \nabla^2 \mathbf{u}_A^*, \tag{1.2.3}$$

$$\frac{\partial \mathbf{u}_W^*}{\partial t^*} + (\mathbf{u}_W^* \cdot \nabla)\,\mathbf{u}_W^* = -\frac{1}{\rho_W}\nabla p_W^* + \nu_W \nabla^2 \mathbf{u}_W^*. \tag{1.2.4}$$

Together they make up the Navier-Stokes equations. Here, $p_{A,W}^*$ is the dimensional pressure within the flow, $\mu_{A,W}$ the dynamic viscosities and $\rho_{A,W}$ the densities upon which the method is based.

If the region of water has length scale $L$ and $U_W$ is a characteristic velocity of the flow within that region, (1.2.3), (1.2.4) may be non-dimensionalised using the following scalings:

$$\mathbf{u}_A^* = U_W \hat{\mathbf{u}}_A, \qquad\qquad\qquad \mathbf{u}_W^* = U_W \hat{\mathbf{u}}_W, \tag{1.2.5}$$

$$p_A^* = \rho_W U_W^2 \hat{p}_A, \qquad\qquad\qquad p_W^* = \rho_W U_W^2 \hat{p}_W, \tag{1.2.6}$$

$$t^* = \frac{L}{U_W}t. \tag{1.2.7}$$

With the variables in this form we may define a dimensionless parameter which characterises

the flow, the Reynolds number:

$$\text{Re}_W = \frac{U_W L}{\nu_W}. \tag{1.2.8}$$

The Reynolds number provides a quantitative measure of the relative importance of inertial and

viscous forces; high Reynolds numbers relate to nominal viscosity or possibly inviscid flows,

low Reynolds numbers to slow or highly viscous flows and so on. This, and the scalings given

above, leave (1.2.3), (1.2.4) as follows,

$$\frac{\partial \hat{\mathbf{u}}_A}{\partial t} + (\hat{\mathbf{u}}_A \cdot \nabla) \hat{\mathbf{u}}_A = - \left( \frac{\rho_W}{\rho_A} \right) \nabla \hat{p}_A + \left( \frac{\nu_A}{\nu_W} \right) \text{Re}_W^{-1} \nabla^2 \hat{\mathbf{u}}_A, \tag{1.2.9}$$

$$\frac{\partial \hat{\mathbf{u}}_W}{\partial t} + (\hat{\mathbf{u}}_W \cdot \nabla) \hat{\mathbf{u}}_W = -\nabla \hat{p}_W + \text{Re}_W^{-1} \nabla^2 \hat{\mathbf{u}}_W. \tag{1.2.10}$$

The continuity equations (1.2.1), (1.2.2) remain unaltered except for the obvious change of the

velocity vector.

Defining our small density ratio parameter $\epsilon = \rho_A / \rho_W$, and assuming $\text{Re}_W, t \ll O(1/\epsilon)$,

$\mu_A / \mu_W = O(\epsilon)$, $\nu_A / \nu_W = O(1)$, we may rescale the non-dimensional velocity and pressure

within the water as

$$(\hat{\mathbf{u}}_W, \hat{p}_W) = \epsilon(\mathbf{u}_W, p_W). \tag{1.2.11}$$

Similarly in the air

$$(\hat{\mathbf{u}}_A, \hat{p}_A) = (\mathbf{u}_A, \epsilon p_A). \tag{1.2.12}$$

The result of this is to make every term in the momentum equations for the surrounding air flow

$O(1)$, reducing the theoretical problem in air to Navier-Stokes flow past a quasi-solid shape, in

view of the nature of the interfacial boundary conditions when the density ratio is small. Thus

we have

$$\frac{\partial \mathbf{u}_A}{\partial t} + (\mathbf{u}_A \cdot \nabla) \mathbf{u}_A = -\nabla p_A + \left( \frac{\nu_A}{\nu_W} \right) \text{Re}_W^{-1} \nabla^2 \mathbf{u}_A, \tag{1.2.13}$$

$$\nabla \cdot \mathbf{u}_A = 0. \tag{1.2.14}$$

Meanwhile, the inertial terms within the water become negligible, leaving the unsteady Stokes-flow equations to be solved in water:

$$\frac{\partial \mathbf{u}_W}{\partial t} = -\nabla p_W + \mathrm{Re}_W^{-1} \nabla^2 \mathbf{u}_W, \tag{1.2.15}$$

$$\nabla \cdot \mathbf{u}_W = 0. \tag{1.2.16}$$

The two systems interact with one another via matched asymptotic expansions of the interfacial conditions, found from the solution in the air and used to drive the flow within the water. As such, the method may be split into three principal steps:

1. The water is replaced with a fixed solid shape and the full Navier-Stokes equations are solved in air over the body. This solution will give the interfacial stress conditions between the fixed solid shape and the air flow.

2. With the boundary between the two fluids remaining fixed, the unsteady Stokes equations may then be solved inside the droplet, subject to the interfacial conditions prescribed at the boundary, and a no slip condition along any real solid walls. This solution yields values for the velocities $\mathbf{u}_W$.

3. With a solution to the flow within the water, a new interface shape at time $t + \delta t$ is in effect determined from the kinematic condition. This can be taken to act to slightly modify the calculation in step one. The process is then repeated for subsequent time steps.

The effects of gravity and surface tension are not taken into account in this model. It would be relatively simple to include these forces via a Youngs-Laplace pressure jump, however, as discussed by [62], for a Weber number, $We = \rho_W U_W L / \sigma$ comparable in size to our small ratios parameter, the effects of both of these may be assumed to be negligible. Here $\sigma$ is surface tension.

Of course, the field of study of two fluid flows is vast (again see [27], [28], [25]). As our

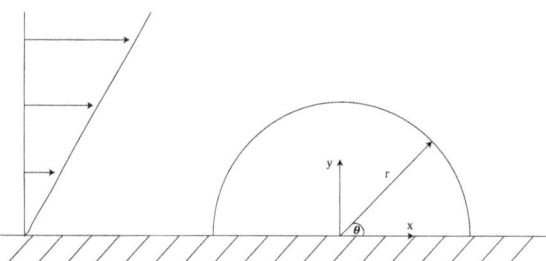

Figure 1.1: A diagram of the flow regime for a modelling the deformation of an initially semi-circular droplet.

title suggests, we limit our investigation to that of water droplets in a surrounding air flow adhering to a solid flat plate or wall. The leading and trailing contact points are pinned down in effect under the present conditions, and we may focus our attention on deformation before rolling or sliding occurs. Far upstream from the droplet, the air flow is unidirectional and has uniform shear, $u \sim y$, for streamwise velocity $u$. A schematic diagram of the flow regime is given graphically in figure (1.1). Many authors have analysed this and similar problems in the past, notable examples including Spelt [64], Dussan V. [15] Dimitrakopoulos [14], but theoretical papers on the present type of nonlinear interaction between the two fluids which is believed to make our study novel - as we shall demonstrate - appear to be few and far between.

A key feature of the paper by Smith and Purvis [62] is the suggestion of two temporal stages to the problem of droplet deformation. First is the early stage where $t$ is of $O(1)$ and within the calculations of the flow in air the droplet shape may be taken as steady to leading order. The air-water interaction of this stage is predominantly one-way. Second, in the later stage, as the $O(\epsilon)$ term in the droplet's shape begins to have considerable influence, the interface becomes severely distorted and the air-water interaction becomes fully non linear. Our work does indeed lead to the identification of two such stages. This problem in particular - the behaviour at larger times - turns out to be rich in fluid dynamical interest and raises a number of intriguing issues as we shall see. Building to this conclusion brings in the analysis of many canonical problems along the way.

## 1.3 Outline of the thesis

In the first part of the thesis we tackle an example of an initially semicircular droplet with $Re_W$ of $O(1)$, as pictured in (1.1). In such a case the small-ratio theory works satisfactorily but applying (within the theory) the full stress conditions along the interface turns out to be computationally expensive and potentially complicated. Instead we choose to take as the interfacial conditions as requirements on pressure and vorticity in order to gain insight. This is contrary to the outline in [62] and a discussion of how this relates to the problem of full stress conditions along the boundary between the fluids is postponed until Chapter 4.

We begin in Chapter 2 by focusing on step one of the method of modeling the interface and droplet deformation: the flow in air over a semicircular fixed solid shape. The lid-driven cavity provides a classical test for our numerical procedure, where a vorticity-streamfunction form of the equations is solved via an iterative finite difference procedure. This forms the basis of the time marching algorithm which itself may be extended and applied to the polar form of the full time dependent Navier-Stokes flow over a semicircle.

With a solution in air over the obstacle we switch our focus to step two in Chapter 3: the unsteady Stokes flow within the droplet. To understand and explore the dynamics of the flow within the droplet alone, we first abandon temporarily the outer solution and prescribe idealised boundary conditions along the free surface. We examine three cases: the so-called circular seed and a square droplet, both of which give insight into behaviour of the fluid in the full problem, before moving to the polar form of Stokes flow within the semicircular droplet itself. Throughout this chapter, a pressure-vorticity form of the Stokes equations is used. For the circular seed, the use of series solutions proves convenient and we show that the velocities within the droplet become steady with time. A series may also be applied to the square droplet as a validation for

a steady state iterative finite difference procedure similar to that seen in the lid driven cavity.

Extension of this to a time marching algorithm suggests that the square droplet - unlike the

circular seed - has velocities which grow linearly with time. This result is mirrored in the time

dependent solution of the semicircular droplet.

Finally the two methods are combined in Chapter 4, so that the time dependent interfacial

conditions found from the external flow field drive the solution within the droplet and deter-

mine the movement of its free surface. In such a case the early and later temporal stages are

associated with $t = O(1)$ and $t = O(\epsilon^{-1/2})$ respectively and an analytical discussion of the

later stage is given, with comparisons of our results to the behaviour postulated in [62]. The

limitations of our model have origins in the interfacial conditions chosen. We present details

of the limiting case in which our results hold and discuss how they relate to a more general case.

Working with a pressure vorticity form of the Stokes equations within the droplet in Chap-

ter 3, the Cauchy-Riemann equations act to enforce a no slip condition along a wall. This

turns out to be an interesting problem in its own right and we investigate the possible solutions

further via a simplified problem in an additional chapter at the end of the first part of the thesis.

The semicircular droplet provides some excellent insight into the problem, but the effect of

severe droplet distortion on the solution in air is difficult to investigate numerically in the later

stage, $t = O(\epsilon^{-1/2})$. We have found a special case however: a droplet contained within the

boundary layer of an external fluid, for which we have successfully modelled the two way

nonlinear interaction between air and water associated with the second temporal stage, with full

stress conditions to leading order along the interface. This is the main motivation of the second

part of the thesis.

In Chapter 6 we spend some time discussing the dynamics of the boundary layer and setting up the triple deck structure required for our problem. Reviews of triple-deck similar and similar theory are given by Smith and Merkin [60], Sychev et al. [69] and Sobey [63]. We present some linear solutions of flow in air over a solid obstacle, which forms the basis of the fully interacting model of droplet deformation of Chapter 8.

The results of Chapter 6 present an opportunity to investigate the steady shapes which an attached droplet within a triple deck structure may take. Applying a lubrication approximation to the second fluid allows us to derive a Reynolds lubrication type expression, which links the shape of the interface to gravity, surface tension and shear forces. We spend some time in Chapter ?? analysing the families of solutions which this expression provides - results which do not depend on a small density assumption.

Returning to the general theme in Chapter 8 we find that the results for air flow over a solid obstacle of Chapter 6 may be extended via a numerical algorithm to provide a fully nonlinear solution. This serves to act as step one in the method of modelling the droplet deformation, while the solution in water sees a second Reynolds lubrication type expression arise. In this case, the properties of the flow in water are entirely eliminated from the problem, and in the special case where the displacement function in air is zero, step two of the method is not required: the shape of the interface may be found directly from the solution in air. This approach is valid for the later temporal stage, $t = O(\epsilon^{-1})$ in this case, as the interface becomes severly distorted. Results of our simulations, and a brief discussion of the finite time break-up of the boundary layer are given there.

# Part I

# A semicircular droplet

## Chapter 2

# Flow in air

As outlined in the previous chapter, the method of modelling droplet deformation which this thesis employs comprises of three stages: a flow in air over a solid shape, a flow in water subject to given interface conditions and a feedback mechanism, where the shape change in the droplet acts to alter the flow in air.

For the current problem we focus our attention on the early stage of deformation of a wall mounted semicircular droplet only, that is, while the shape remains semicircular to leading order. Doing so allows us to neglect the final step and hence to treat the flow in air as an isolated problem. That is the task of this chapter, to find a solution to the Navier-Stokes equations in air (1.2.13), (1.2.14) over a solid semicircle attached to a wall. We repeat the diagram of the flow regime here in figure (2.1).

With the origin defined at the center of the semicircle, the equations and boundary conditions are simplest in polar form. The wall of the semicircle lies along $r = 1$ and $\theta = \pi$ refers to the leading edge of the droplet. This polar setup is given graphically in figure (2.2a).

Once in this form, the similarities between the flow over a semicircular body attached to a wall and the model problem of a driven cavity, shown in figure (2.2b), become clear. Applying

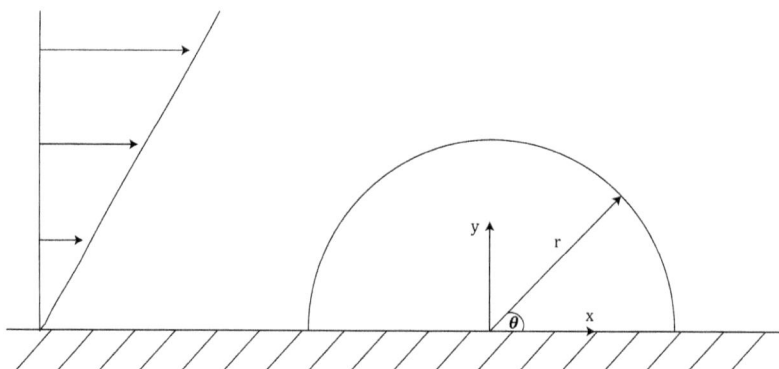

Figure 2.1: A schematic diagram of the flow in air over a semicircular surface mounted obstacle.

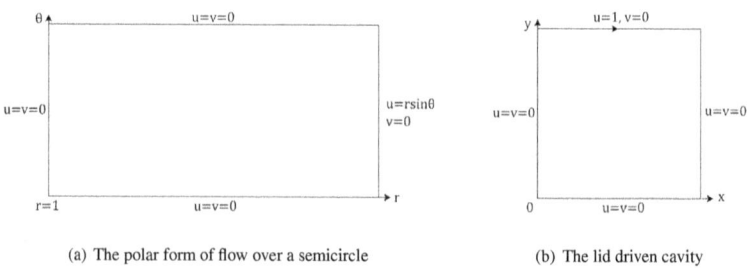

(a) The polar form of flow over a semicircle

(b) The lid driven cavity

Figure 2.2: Comparison between polar form of flow over a semicircular body and the lid driven cavity.

(1.2.13), (1.2.14) to the flow within the cavity, the only key difference between the two systems

is the farfield velocity condition. Because of these similarities, the simplicity of the geometry

and the abundance of literature on the example, the lid driven cavity of Burggraf [9] and many

others, serves as an excellent test for our method outside of the droplet. We choose then to first

tackle the lid driven cavity problem, before extending our method to the full polar system of

flow over a semicircle. We spend some time in the following section detailing the methods used

in our numerical procedure, since they are repeated throughout the chapters in this part of the

thesis.

## 2.1 The lid driven cavity

The fluid here is contained within a unit square with vertices at $(0,0)$, $(0,1)$, $(1,0)$ and $(1,1)$.

Three of the walls are stationary so that the no slip condition $u = v = 0$ applies. Along the

top, or lid, the flow is driven by a moving wall of velocity $u = 1$ in the positive $x$ direction.

The non dimensional Navier-Stokes equations with application of the small ratios, $(\hat{u}_A, \hat{p}_A) =$

$(u_A, \epsilon p_A)$, govern the flow within the cavity. These may be found in the previous chapter

(1.2.13),(1.2.14), but are repeated here for clarity:

$$\frac{\partial \mathbf{u}_A}{\partial t} + (\mathbf{u}_A \cdot \nabla)\mathbf{u}_A = -\nabla p_A + \left(\frac{\nu_A}{\nu_W}\right)\mathrm{Re}_W^{-1}\nabla^2\mathbf{u}_A. \tag{2.1.1}$$

$$\nabla \cdot \mathbf{u}_A = 0. \tag{2.1.2}$$

Equations (2.1.1) and (2.1.2) may be written as a vorticity-streamfunction system:

$$\nabla^2\zeta = \mathrm{Re}_W\left(\frac{\nu_A}{\nu_W}\right)\left(\frac{\partial \psi}{\partial y}\frac{\partial \zeta}{\partial x} - \frac{\partial \psi}{\partial x}\frac{\partial \zeta}{\partial y} + \frac{\partial \zeta}{\partial t}\right), \tag{2.1.3}$$

$$\nabla^2\psi = -\zeta, \tag{2.1.4}$$

where the vorticity $\zeta$ and the streamfunction $\psi$ are defined as

$$\zeta = \frac{\partial v}{\partial x} - \frac{\partial u}{\partial y}, \qquad u = \frac{\partial \psi}{\partial y}, \qquad v = -\frac{\partial \psi}{\partial x}. \tag{2.1.5}$$

For simplicity we rewrite

$$\widehat{\mathrm{Re}} = \left(\frac{\nu_W}{\nu_A}\right)\mathrm{Re}_W, \tag{2.1.6}$$

throughout the remainder of this chapter. For a small dynamic viscosity ratio, $(\nu_A/\nu_W)$ will be of $O(1)$, and is around $1/55$ in the case of air and water.

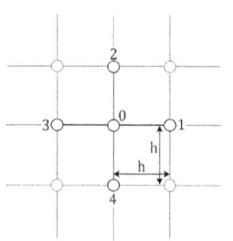

Figure 2.3: The grid labelling system.

The numerical method begins by discretising the flow domain into a uniform rectangular grid of spacing $h$. The labelling system for the grid points and their immediate neighbors is shown in the five-point stencil in figure (2.3), and referred to by a subscript from now on. With this setup, we may apply a second order accurate finite difference procedure by replacing the derivatives in the system of equations (2.1.4), (2.1.3) and so on, with their difference quotients in the standard way. For example,

$$\left.\frac{\partial\psi}{\partial x}\right|_{(x_0,y_0)} = \frac{\psi_1 - \psi_3}{2h}, \qquad \left.\frac{\partial^2\psi}{\partial x^2}\right|_{(x_0,y_0)} = \frac{\psi_1 - 2\psi_0 + \psi_3}{h^2} \qquad (2.1.7)$$

$$\left.\frac{\partial\psi}{\partial y}\right|_{(x_0,y_0)} = \frac{\psi_2 - \psi_4}{2h}, \qquad \left.\frac{\partial^2\psi}{\partial y^2}\right|_{(x_0,y_0)} = \frac{\psi_2 - 2\psi_0 + \psi_4}{h^2}. \qquad (2.1.8)$$

After some rearranging, this substitution (2.1.7), (2.1.8) leaves (2.1.4) as

$$\psi_1 + \psi_2 + \psi_3 + \psi_4 - 4\psi_0 = -h^2\zeta_0. \qquad (2.1.9)$$

The form of the expression for vorticity (2.1.3) differs for a steady state or time dependent problem and so is postponed until the next sections. It remains to determine the boundary conditions in the appropriate form before (2.1.9) may be investigated.

The streamfunction is taken to be zero along the four walls, but the value of the vorticity along the boundaries must be derived from (2.1.9) using a method given in Woods [75] and Dennis [13] as follows.

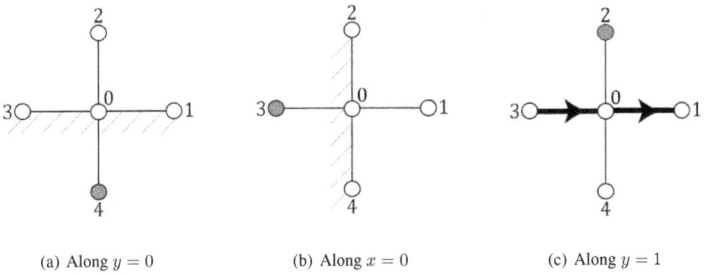

(a) Along $y = 0$              (b) Along $x = 0$              (c) Along $y = 1$

Figure 2.4: The grid points along the boundaries in the driven cavity

Figure (2.4a) shows a point on the discretised grid lying along $y = 0$ and its four neighbors. Point 4 lies outside of the cavity, and so we aim to eliminate $\psi_4$ from (2.1.9). We do so by considering the finite difference form of (2.1.5) given that the no slip condition dictates that $u = v = 0$. This suggests that the shadow point $\psi_4$ takes the value

$$\psi_4 = \psi_2 + O(h^3). \tag{2.1.10}$$

Further, since $\psi$ is zero along a wall, $\psi_0, \psi_1, \psi_3 = 0$ which substituted into (2.1.9) leads to the second order accurate boundary condition for the vorticity along $y = 0$,

$$\zeta_b = -\frac{2\psi_i}{h^2}. \tag{2.1.11}$$

Here subscripts $b$ and $i$ refer to the grid points on the boundary and immediately interior to the boundary respectively. Along $y = 0$ for example, $b = 0$ and $i = 2$. The argument and result are the same along $x = 0$ and $x = 1$, except here $\psi_0, \psi_2, \psi_4 = 0$ and $\psi_3 = \psi_1$ (see figure (2.4b)).

Along $y = 1$, shown in figure (2.4c), we again have $\psi_0, \psi_1, \psi_3 = 0$ but now the wall is moving with speed $u = 1$. The finite difference form of (2.1.5) becomes

$$\frac{\partial \psi}{\partial y} = \frac{\psi_2 - \psi_4}{2h} = 1, \tag{2.1.12}$$

which allows us to eliminate the shadow point $\psi_2$ from (2.1.9) and makes the boundary condition for vorticity along $y = 1$

$$\zeta_b = -\frac{2\psi_i}{h^2} - \frac{2}{h}. \tag{2.1.13}$$

### 2.1.1 The steady state problem

It is believed that the driven cavity has unique steady laminar solutions for a given Reynolds number, over some range at least. Whilst we aim to achieve a full time dependent solution, tackling the steady state problem separately will allow us to validate our time marching results and method, since we expect it to converge to this unique steady state result. The steady state problem has identical boundary conditions and field equations except in (2.1.3) where the time derivative term is dropped. Using the same labelling system and difference quotients (2.1.7)-(2.1.8) to derive the finite difference form for the field equation for $\zeta$ then, the full set of equations and boundary conditions for the steady state problem is as follows:

$$\zeta_0 = \frac{1}{4}(\zeta_1 + \zeta_2 + \zeta_3 + \zeta_4) - \frac{\widehat{\mathrm{Re}}}{16}(\psi_2 - \psi_4)(\zeta_1 - \zeta_3) + \frac{\widehat{\mathrm{Re}}}{16}(\psi_1 - \psi_3)(\zeta_2 - \zeta_4), \tag{2.1.14}$$

$$\zeta_b = -\frac{2\psi_i}{h^2} \quad \text{on} \quad x = 0, 1 \quad y = 0, \tag{2.1.15}$$

$$\zeta_b = -\frac{2\psi_i}{h^2} - \frac{2}{h} \quad \text{on} \quad y = 1, \tag{2.1.16}$$

$$\psi_0 = \frac{\psi_1 + \psi_2 + \psi_3 + \psi_4 + h^2\zeta}{4}, \tag{2.1.17}$$

$$\psi = 0 \quad \text{on} \quad x = 0, 1 \quad y = 0, 1. \tag{2.1.18}$$

A simple iterative process may be used to solve this coupled system: (2.1.14), with boundary conditions (2.1.15), (2.1.16); and (2.1.17) with boundary conditions (2.1.18). First, an initial guess of $\zeta = 0$ everywhere is made. Based on this guess (2.1.18) is solved by sweeping across the domain from left to right, updating the value of the streamfunction at each point in turn. The solution $\psi$ is then used in (2.1.15), (2.1.16), to determine the latest boundary conditions for the vorticity, and in (2.1.14), which is solved in the same way as (2.1.18). The result $\zeta$ signifies the beginning of the next iteration and the process is repeated until successive values of both $\zeta$

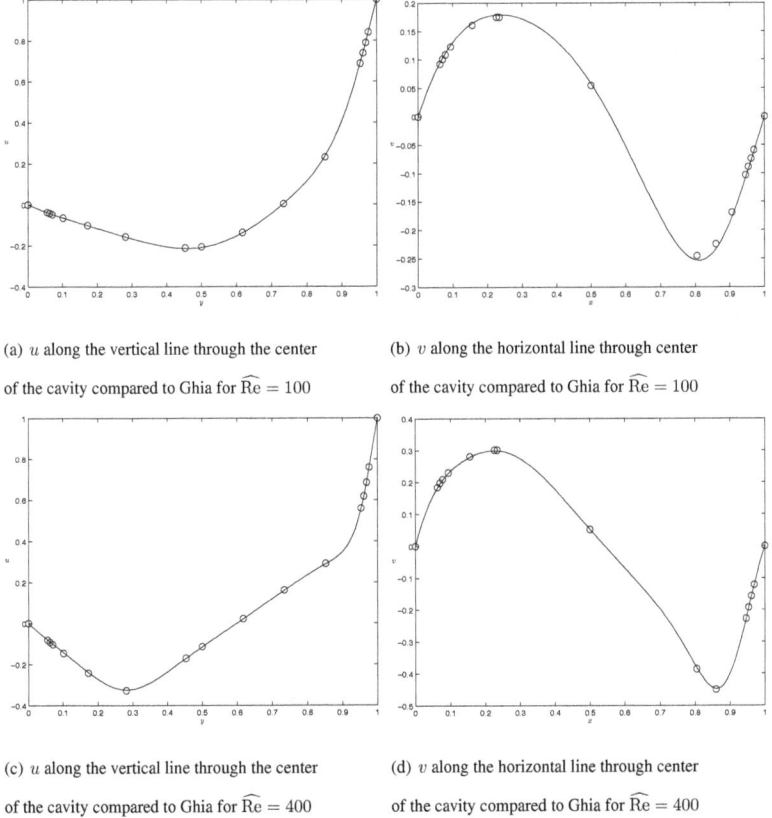

(a) $u$ along the vertical line through the center
of the cavity compared to Ghia for $\widehat{Re} = 100$

(b) $v$ along the horizontal line through center
of the cavity compared to Ghia for $\widehat{Re} = 100$

(c) $u$ along the vertical line through the center
of the cavity compared to Ghia for $\widehat{Re} = 400$

(d) $v$ along the horizontal line through center
of the cavity compared to Ghia for $\widehat{Re} = 400$

Figure 2.5: Comparison between our results and those from Ghia et al. for velocities $u$ and $v$ through the center of the cavity.

and $\psi$ converge. If *old* refers to the value of $\zeta$ at the previous iteration, the system is terminated once

$$\frac{\max|\zeta - \zeta^{old}|}{|\text{mean}(\zeta)|} \leq 10^{-5}, \tag{2.1.19}$$

likewise for $\psi$. We use a grid of 129x129 points and a range of Reynolds numbers from $\widehat{\text{Re}} = 0$ to $\widehat{\text{Re}} = 400$ to take advantage of the abundance of literature on the medium Reynolds number case, while bearing in mind that our main interest is in extending the driven cavity example to the polar form of Navier-Stokes flow over a semicircle at low Reynolds numbers. While higher Reynolds numbers than considered here have boundary layers forming near the walls and can cause singularities at the corner points (in particular at $(x, y) = (0, 1), (1, 1)$ where the velocity is discontinuous) the simple approach of a straightforward finite difference method seems to work well for low to medium Reynolds numbers $\widehat{\text{Re}} \leq 400$: the corner values themselves are not used in the iterative process.

A particularly helpful set of data to use as a comparison is that given in Ghia, Ghia and Shin [18], since it includes tabular results for various Reynolds numbers. Figure (2.5a) and (2.5c) show our results for the $u$ velocity along the vertical line through the geometric center of the cavity for $\widehat{\text{Re}} = 100$ and $\widehat{\text{Re}} = 400$ respectively. Ghia et al.'s data is also plotted and shown as circles. Figures (2.5b) and (2.5d) show the same but for the $v$ velocity through the horizontal line through the center. There appears to be close agreement between the two sets, particularly in the figures for $u$ velocity.

Also shown is a side by side comparison of our results and Ghia et al.'s for the streamfunction and vorticity within the cavity for $\widehat{\text{Re}} = 100$ and $\widehat{\text{Re}} = 400$. These contour plots are shown in

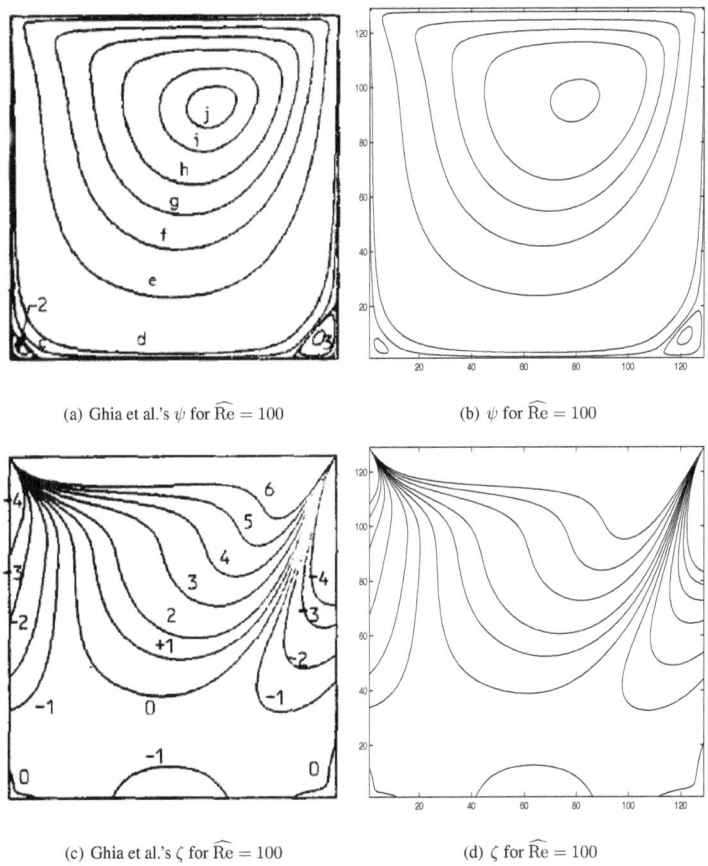

(a) Ghia et al.'s $\psi$ for $\widehat{\mathrm{Re}} = 100$    (b) $\psi$ for $\widehat{\mathrm{Re}} = 100$

(c) Ghia et al.'s $\zeta$ for $\widehat{\mathrm{Re}} = 100$    (d) $\zeta$ for $\widehat{\mathrm{Re}} = 100$

Figure 2.6: Streamfunction and vorticity results comparison for $\widehat{\mathrm{Re}} = 100$.

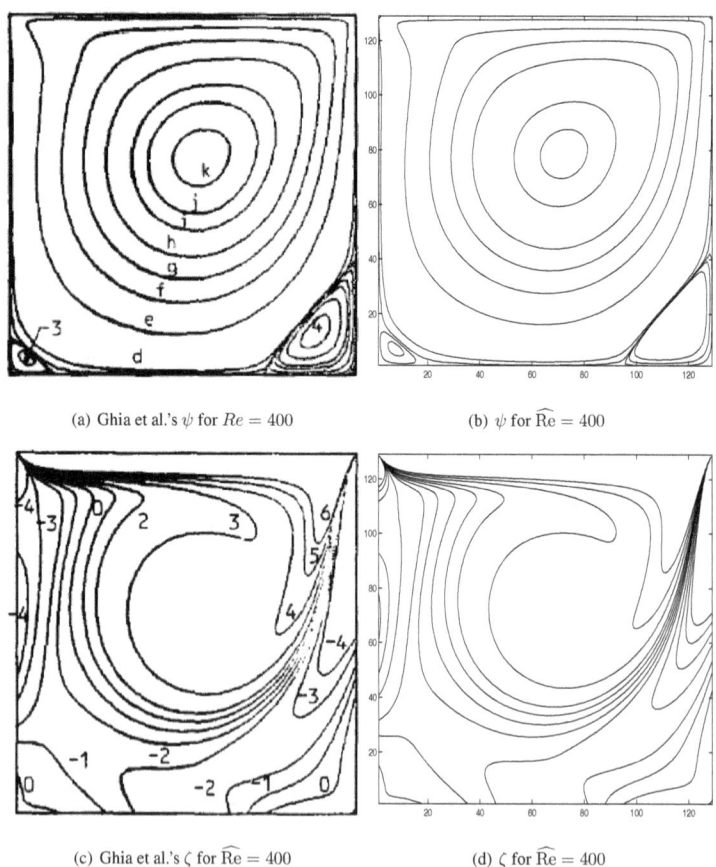

(a) Ghia et al.'s $\psi$ for $Re = 400$           (b) $\psi$ for $\widehat{Re} = 400$

(c) Ghia et al.'s $\zeta$ for $\widehat{Re} = 400$        (d) $\zeta$ for $\widehat{Re} = 400$

Figure 2.7: Streamfunction and vorticity results comparison for $\widehat{Re} = 400$.

figures (2.6) and (2.7). The contours in the streamfunction and vorticity images are taken at

$$\psi = (-0.11, -0.10, -0.90, -0.07, -0.05, -0.03,$$

$$-0.01, -10^{-4}, -10^{-5}, 10^{-6}, 10^{-5}, 5\text{x}10^{-5}), \tag{2.1.20}$$

$$\zeta = (-4, -3, -2, -1, -0.5, 0, 0.5, 1, 2, 3, 4, 5), \tag{2.1.21}$$

as in Ghia et al.'s paper. The two are not identical, but are very close and the trend seems the same in both sets, with the primary vortex occurring near the center of the cavity but offset to the top right corner for $\widehat{Re} = 100$, and smaller recirculation zones appearing in the bottom corners which are larger in the $\widehat{Re} = 400$ image. These differences may be due to graphics interpolation effects.

For completeness, we include contour plots of the steady state vorticity and streamfunction for a range of smaller Reynolds numbers. These are shown in figures (2.8) and (2.9). As one would expect, the vorticity contours and streamlines are symmetrical about the vertical centre-line of the cavity for $\widehat{Re} = 0$. The two eddies in the bottom corners are similar in size. As the Reynolds number increases to around $\widehat{Re} = 50$ the primary vortex moves to the top right hand corner, and we begin to see growth in the eddy in the right hand corner. At higher Reynolds numbers still, the primary vortex begins to move back toward the center of the cavity. We also see the concentration of the vorticity contours increasing, which indicates areas of higher gradients. These results are found on a grid of 129x129 points using the same contours listed above. All this agrees with others' results and suggests our numerical algorithm is working well.

We also found that refining the grid and changing the initial guess for $\zeta$ had little or no effect on the results. Given the close agreement with Ghia et al., we did not think it necessary to include these figures here.

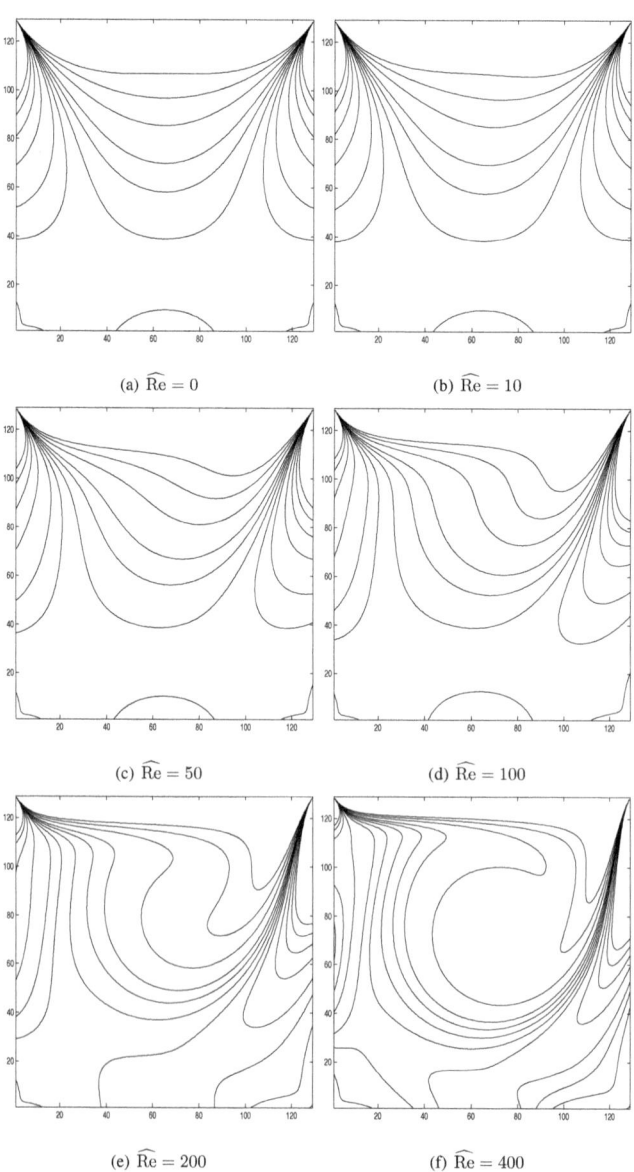

(a) $\widehat{\mathrm{Re}} = 0$                               (b) $\widehat{\mathrm{Re}} = 10$

(c) $\widehat{\mathrm{Re}} = 50$                              (d) $\widehat{\mathrm{Re}} = 100$

(e) $\widehat{\mathrm{Re}} = 200$                             (f) $\widehat{\mathrm{Re}} = 400$

Figure 2.8: Results for $\zeta$ within the lid driven cavity for 129x129 grid points and a range of Reynolds numbers.

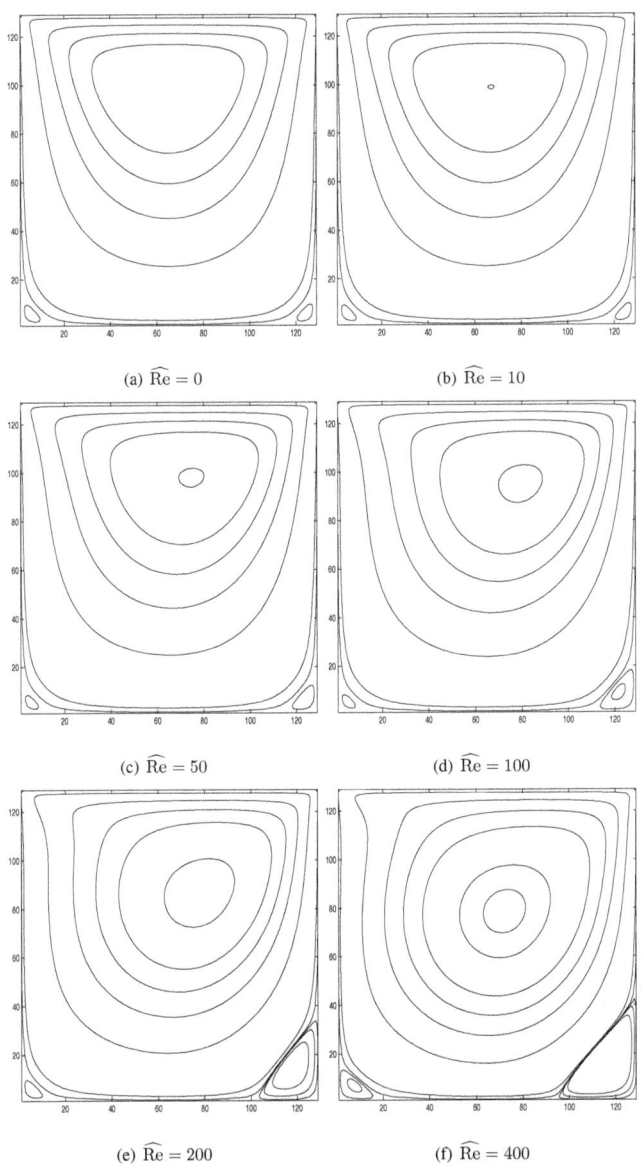

(a) $\widehat{\text{Re}} = 0$

(b) $\widehat{\text{Re}} = 10$

(c) $\widehat{\text{Re}} = 50$

(d) $\widehat{\text{Re}} = 100$

(e) $\widehat{\text{Re}} = 200$

(f) $\widehat{\text{Re}} = 400$

Figure 2.9: Results for $\psi$ within the lid driven cavity for 129x129 grid points and a range of Reynolds numbers.

### 2.1.2   The time dependent problem

Comparisons to Ghia et al.'s paper give us some confidence that our numerical procedure is working well and so we switch our focus to extending the algorithm to the full time dependent problem. The equations and boundary conditions remain the same, namely (2.1.15), (2.1.16) and (2.1.18), except (2.1.14) is replaced by

$$\zeta_0^k = \frac{\delta t \left(\zeta_1^k + \zeta_2^k + \zeta_3^k + \zeta_4^k\right)}{4\delta t + h^2 \widehat{\mathrm{Re}}} + \frac{\delta t \widehat{\mathrm{Re}} \left\{ (\psi_1^k - \psi_3^k)(\zeta_2^k - \zeta_4^k) - (\psi_2^k - \psi_4^k)(\zeta_1^k - \zeta_3^k) \right\}}{4(4\delta t + h^2 \widehat{\mathrm{Re}})}$$
$$+ \frac{h^2 \widehat{\mathrm{Re}} \widehat{\zeta_0^{k-1}}}{4\delta t + h^2 \widehat{\mathrm{Re}}}. \tag{2.1.22}$$

Within (2.1.22) superscript $k$ refers to the latest value at the current time step, and $k - 1$ to the converged solution at the previous time step. We begin with an initial condition of $\zeta^{k=0} = 0$ and use the same iterative process as for the steady state to find $\psi^{k=1}$ and $\zeta^{k=1}$. From here, we may solve for the next time step, using $\zeta^{k=1}$ to find $\zeta^{k=2}$, $\psi^{k=2}$ and so on. A time step of $\delta t = \widehat{\mathrm{Re}} h^2/2$ was used in all calculations.

The time marching results for the vorticity contours when $\widehat{\mathrm{Re}} = 100$ are shown in figure (2.10), and for the streamlines in figure (2.11). We see that the primary vortex forms near the top right hand corner of the cavity and moves toward the center as time increases, where it stays. Two smaller vortices form at the bottom corners and remain there for all time. In both figures the steady state results (2.10f), (2.11f) discussed previously are also shown for comparison with the $t = 16$ results (2.10e), (2.11e). We see that the time marching method does indeed converge to the steady state solution, which increases confidence in the present numerical procedure.

## 2.2   Air flow over a surface mounted semicircle

With reasonable results for our time marching algorithm and the steady state solution matching closely to those of Ghia et al. we now look to apply the Navier-Stokes solver to the polar form of shear flow over a semicircular body attached to a wall. The same definition of $\widehat{\mathrm{Re}}$ seen in (2.1.6) is used, and when this parameter is $O(1)$, the setup corresponds to step one in the

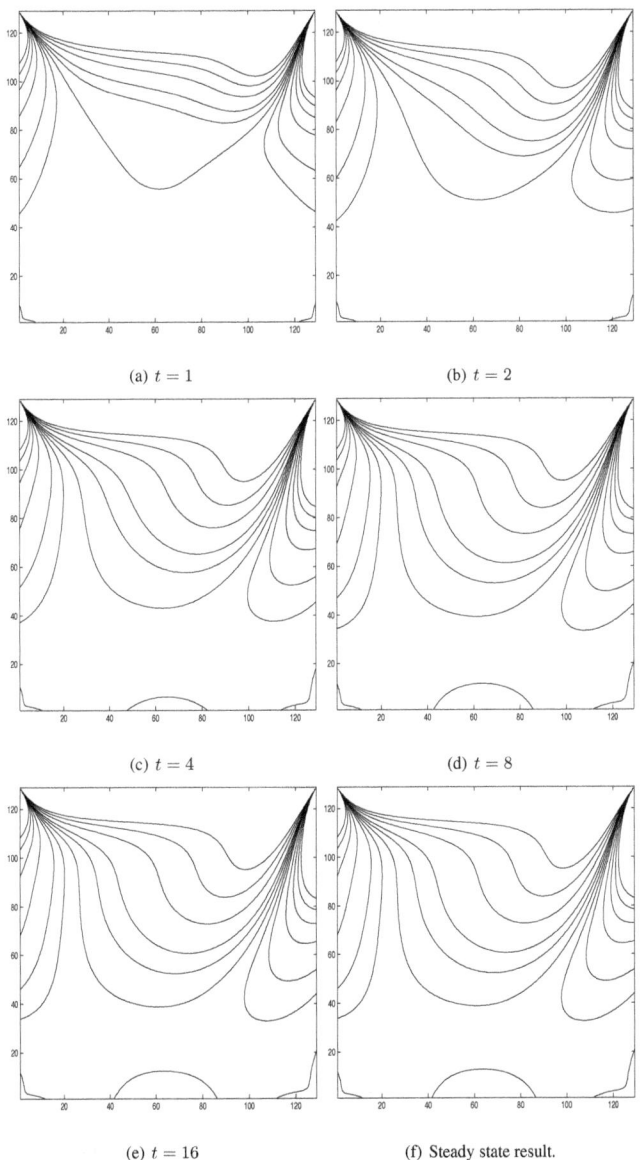

(a) $t = 1$

(b) $t = 2$

(c) $t = 4$

(d) $t = 8$

(e) $t = 16$

(f) Steady state result.

Figure 2.10: Unsteady results for $\zeta$ within the lid driven cavity for 129x129 grid points, and $\widehat{\text{Re}} = 100$. The corresponding steady state result (f) is included for comparison.

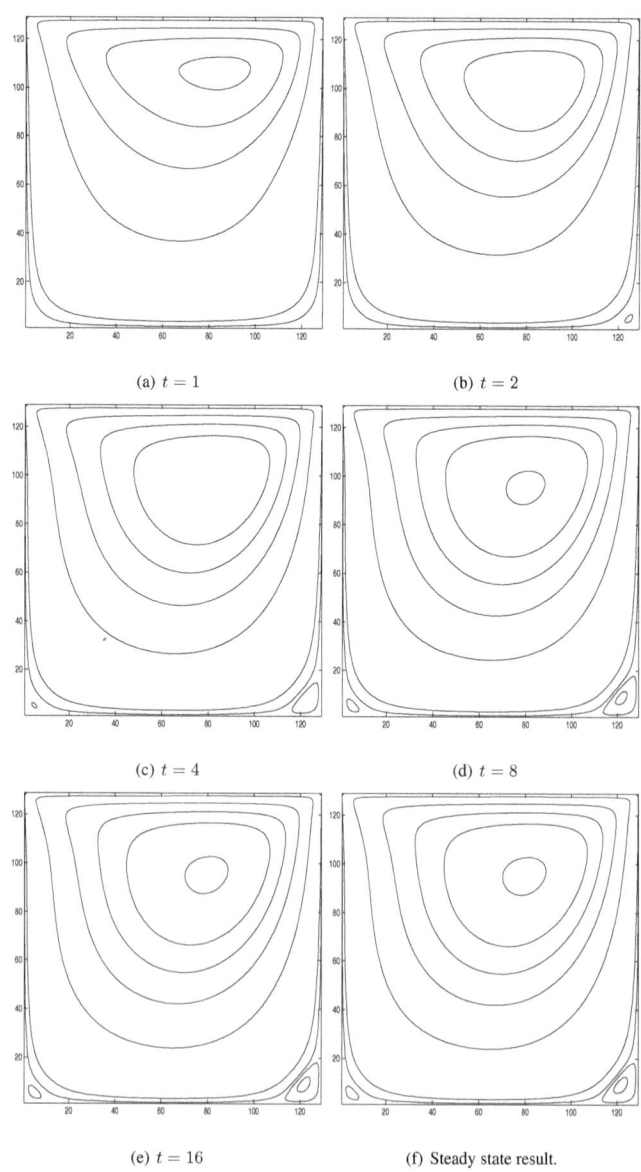

(a) $t = 1$

(b) $t = 2$

(c) $t = 4$

(d) $t = 8$

(e) $t = 16$

(f) Steady state result.

Figure 2.11: Unsteady results for $\psi$ within the lid driven cavity for 129x129 grid points, and $\widehat{\mathrm{Re}} = 100$. The corresponding steady state result (f) is included for comparison.

method for modelling droplet deformation outlined in Chapter 1. Our interest in this problem

lies in numerically determining the pressure and vorticity along the surface of the obstacle, with

the aim of using the solution to drive the flow within the water droplet, seen in Chapter 4.

The flow regime is given graphically in Cartesian form at the beginning of the chapter in

figure (1.1), and in polar form in figure (2.2a). The semicircular hump has center $(0,0)$ and

radius 1 so that the outer wall of the body lies at $r = 1$. The air flows generally from left to

right and far from the body is of uniform shear $u \sim y$.

Following the approach taken in the lid driven cavity problem, after application of the small

ratios scaling, we consider the Navier-Stokes equations in vorticity-streamfunction form:

$$\nabla^2 \zeta = \widehat{\mathrm{Re}} \left( \frac{1}{r} \frac{\partial \psi}{\partial \theta} \frac{\partial \zeta}{\partial r} - \frac{1}{r} \frac{\partial \psi}{\partial r} \frac{\partial \zeta}{\partial \theta} + \frac{\partial \zeta}{\partial t} \right), \qquad (2.2.1)$$

$$\nabla^2 \psi = -\zeta, \qquad (2.2.2)$$

where streamfunction and vorticity are defined as

$$\zeta = \frac{1}{r} \frac{\partial}{\partial r} \left( r u_\theta \right) - \frac{1}{r} \frac{\partial}{\partial \theta} \left( u_r \right), \qquad (2.2.3)$$

$$\frac{1}{r} \frac{\partial \psi}{\partial \theta} = u_r, \qquad \frac{\partial \psi}{\partial r} = -u_\theta. \qquad (2.2.4)$$

for radial and azimuthal velocities $u_r$, $u_\theta$.

To allow for comparisons between the previous example and this problem, we base our numer-

ical algorithm on that seen in the lid driven cavity, where a uniform discretisation was applied.

It seems sensible to do the same here also. With this approach, to ensure the numerical grid has

points which lie exactly along the solid wall at $\theta = 0$ and $\theta = \pi$, we must restrict our choice

of grid spacing to be a fraction of $\pi$. A uniform discretisation of the rectangular region $(r, \theta)$,

places an additional restriction on the value of the Reynolds number we may practically use

in our computations and this is discussed later. Despite these restrictions, applying a constant grid spacing throughout the geometry appears to work well for our purpose as we shall see, and

has the advantage of allowing us to express the finite difference version of the equations and

boundary conditions in a simple form.

To derive the boundary conditions we begin with the no slip along walls where the fluid is

attached, so that $u_r$, $u_\theta$ and $\psi$ are all zero along $\theta = 0, \pi$ and $r = 1$. For the vorticity here,

we follow an identical argument to that in the previous example: the second order accurate

finite difference form of (2.2.4) is used to eliminate the shadow points in the finite difference

form of (2.2.2). Again, subscripts $b$ and $i$ are used to refer to the grid points on the boundary

and immediately interior to the boundary respectively. In the present case, these boundary

conditions take the form:

$$\psi = 0, \quad \zeta_i = \frac{-2\psi_i}{r_b^2 h^2} \quad \text{along } \theta = 0, \pi, \qquad (2.2.5)$$

$$\psi = 0, \quad \zeta_i = \frac{-2\psi_i}{h^2} \quad \text{along } r = 1. \qquad (2.2.6)$$

The fourth condition for both $\zeta$ and $\psi$ as $r \to \infty$, differs from the driven cavity, where a moving

wall at a fixed distance gave a uniform velocity. For the current problem, a uniform far field

shear $u \sim y$ directly yields a condition for $\psi$, which, when combined with (2.2.2) provides the

final condition for $\zeta$:

$$\psi = \frac{1}{2}r^2 \sin^2 \theta, \quad \zeta = -1 \text{ as } r \to \infty. \qquad (2.2.7)$$

It is this far field condition (2.2.7) which causes the computational restriction on the maximum

Reynolds number we can use, a problem we did not encounter in the lid driven cavity. While the

equations assume the setup to be semi infinite in $y$, the computational algorithm takes a fixed

boundary at a finite $r$ value, $r_\infty$. The choice of $r_\infty$ in the numerics must be sufficiently large

that any recirculatory motion downstream of the obstacle does not cause an upstream effect via

the $r_\infty$ boundary. As the Reynolds number increases, so too does the length of such an eddy

and so must the minimum usable value of $r_\infty$. With a uniform grid spacing, a larger choice of

$r_\infty$ greatly increases the number of grid points used and becomes computationally expensive.

If our interest lay in flows with larger Reynolds numbers this could easily be avoided by using a

multi-grid method. For our purpose however, where $\widehat{Re} = O(1)$, this simple setup of a uniform

grid works well for a choice of $r_\infty = 15$ and $Re <= 5$. No evidence of Stokes' paradox was

seen with choices of $\widehat{Re}$ within this range.

### 2.2.1   The steady state problem

In a similar way to the lid driven cavity problem, it is believed that the flow over a semicircle

may converge in time to a steady laminar solution, unique for a given Reynolds number. If this

is the case, it seems sensible to once again begin by tackling the steady state problem, as it will

allow us later to validate our time marching algorithm.

In the steady state problem, the time derivative term in (2.2.1) is neglected. A second or-

der accurate finite difference procedure is applied to (2.2.2) and (2.2.1) using same labelling

system outlined in the previous section. This leads to the equations which govern the numerical

algorithm:

$$\psi_0 = \left\{ r_0 \left( r_0 + \frac{h}{2} \right) \psi_1 + r_0 \left( r_0 - \frac{h}{2} \right) \psi_3 + \psi_2 + \psi_4 + r_0^2 h^2 \zeta_0 \right\} / 2(1 + r_0^2), \qquad (2.2.8)$$

$$\zeta_0 = \left\{ r_0 \left( r_0 + \frac{h}{2} \right) \zeta_1 + r_0 \left( r_0 - \frac{h}{2} \right) \zeta_3 + \zeta_2 + \zeta_4 + \right.$$
$$\left. \frac{r_0 \widehat{Re}}{4} (\psi_1 - \psi_3)(\zeta_2 - \zeta_4) - \frac{r_0 \widehat{Re}}{4} (\psi_2 - \psi_4)(\zeta_1 - \zeta_3) \right\} / 2(r_0^2 + 1). \qquad (2.2.9)$$

With an initial guess of $\zeta = 0$ everywhere, an identical iterative process as detailed in section

(2.1.1) is used to solve the two coupled systems (2.2.8) and (2.2.9) subject to the boundary con-

ditions (2.2.5), (2.2.6) and (2.2.7): the streamfunction and vorticity are solved simultaneously,

taken as known at each iteration and then updated accordingly. In all of the calculations, the

system was terminated once the error between successive iterations of $\zeta$ and $\psi$ reach below

$10^{-5}$. It was found that no relaxation was required for the Reynolds numbers considered.

Results for the streamfunction at $\widehat{\mathrm{Re}} = (0, 1, 3, 5)$, calculated on a grid with uniform spacing $h = \pi/150$ are shown in figure (2.12). In these images, contours are plotted at

$$\psi = \left(-10^{-4}, -10^{-3}, -10^{-2}, -10^{-1}, 0, 10^{-3}, 10^{-2}, 10^{-1}, 0.2, 0.3, 0.5,\right.$$

$$\left.0.7, 0.9, 1.2, 1.6, 2, 2.3, 2.6, 3.5, 5, 6\right). \tag{2.2.10}$$

For clarity, it was decided to not include contour labels within the plots and instead a colour bar is shown at the top of the figure. The images demonstrate that the streamline pattern is symmetrical for $\widehat{\mathrm{Re}} = 0$ with two small recirculation zones on either side of the obstacle. Similar eddies were seen in the lid driven cavity example and we know from Moffatt [37] to expect eddies to exist near a corner between two intersecting planes where the angle is less than $146.3°$, as it is in our problem. The length of the downstream eddy appears to almost increase linearly with Reynolds number, which agrees with the findings of Bhattacharyya, Dennis and Smith [6], although their work extends to much higher Reynolds numbers. We also see a dividing streamline which separates the eddies from the main body of the flow and pinching of the streamlines above the semicircle, both being features which are described by Higdon [23] whose paper included some work on Stokes flow past surface mounted circular crests.

Figure (2.13) shows the results for vorticity, found using the same Reynolds numbers and grid spacing as the streamfunction. A colour bar is also shown at the top of the figure and contours are taken at intervals of $0.1$ between $\zeta = -3$ and $\zeta = 0.1$. In all of the images, $\zeta = -1.3$ corresponds to the outermost contour in the loop immediately above the semi circle, and $\zeta = -0.9$ is the outermost contour both upstream and downstream to the body. To make the image clearer, vorticity values between $-1.3$ and $-0.9$ were not plotted, although these would lie between the two contours described above. Again, the pattern is symmetrical for $\widehat{\mathrm{Re}} = 0$ and the distortion increases with Reynolds number. The general trend of the lines of constant vorticity is consistent with Keller and Takami [29] whose results on flow past a cylinder at low

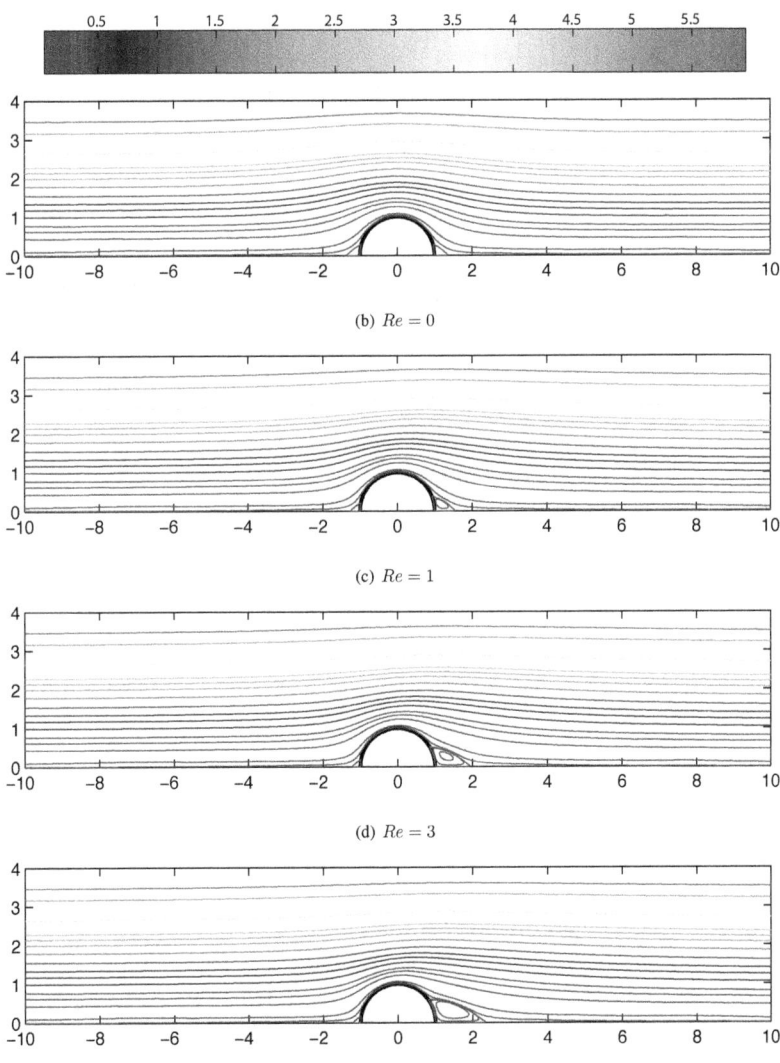

(b) $Re = 0$

(c) $Re = 1$

(d) $Re = 3$

(e) $Re = 5$

Figure 2.12: Streamlines for the steady flow in air over a semicircle with $h = \dfrac{\pi}{150}$ and a range of Reynolds number.

Reynolds numbers were reprinted in Batchelor's book [4].

We also include figures (2.14) and (2.15) to demonstrate the effect of grid refinement on streamfunction and vorticity respectively. In these images, $\widehat{\mathrm{Re}} = 5$ and the grid spacings $\pi/100, \pi/150$ and $\pi/200$ are used. Increasing the number of grid points in the computation appears to have a minimal effect on the results for streamfunction, particularly if we compare the two images with finer grids, (2.14c) and (2.14d). We notice that the vorticity seems much more sensitive to grid spacing as expected, particularly downstream of the semicircle. Overall however, the trend is the same throughout which gives us some confidence that the algorithm is working well. It remains to extend this steady state algorithm to the time dependent problem, and for that purpose we conclude that a grid spacing of around $\pi/150$ is sufficient to produce accurate results without being too computationally expensive.

### 2.2.2 The time dependent problem

To achieve this unsteady numerical solution, we reintroduce the time derivative term in (2.2.1). Using the same labelling system as before, and $k$ to refer to the value at the current time step, we apply a second order accurate finite difference procedure in space, first order accurate backwards finite difference in time, to arrive at the following:

$$\zeta_0^k = \left\{ r_0 \left( r_0 + \frac{h}{2} \right) \zeta_1^k + r_0 \left( r_0 - \frac{h}{2} \right) \zeta_3^k + \zeta_2^k + \zeta_4^k + \frac{r_0 \widehat{\mathrm{Re}}}{4}(\psi_1^k - \psi_3^k)(\zeta_2^k - \zeta_4^k) \right.$$
$$\left. - \frac{r_0 \widehat{\mathrm{Re}}}{4}(\psi_2^k - \psi_4^k)(\zeta_1^k - \zeta_3^k) + \frac{r_0^2 h^2 \widehat{\mathrm{Re}}}{\delta t} \zeta_0^{k-1} \right\} / \left( 2 r_0^1 + 2 + \frac{r_0^2 h^2 \widehat{\mathrm{Re}}}{\delta t} \right) \qquad (2.2.11)$$

This and (2.2.8) are the governing equations for time marching algorithm. The boundary conditions along any solid walls (2.2.5), (2.2.6) remain the same, but impulsively starting a shear flow along $r_\infty$ by using (2.2.7) in an initially quiescent fluid was found to cause an undesired thin layer of vorticity which forms along $r_\infty$ and spreads through the flow field with thickness of order $t^{1/2}$. This feature, known as a Rayleigh layer, occurs when flow at rest is subjected to shear forces along a boundary. The particles are dragged along with the boundary, generating

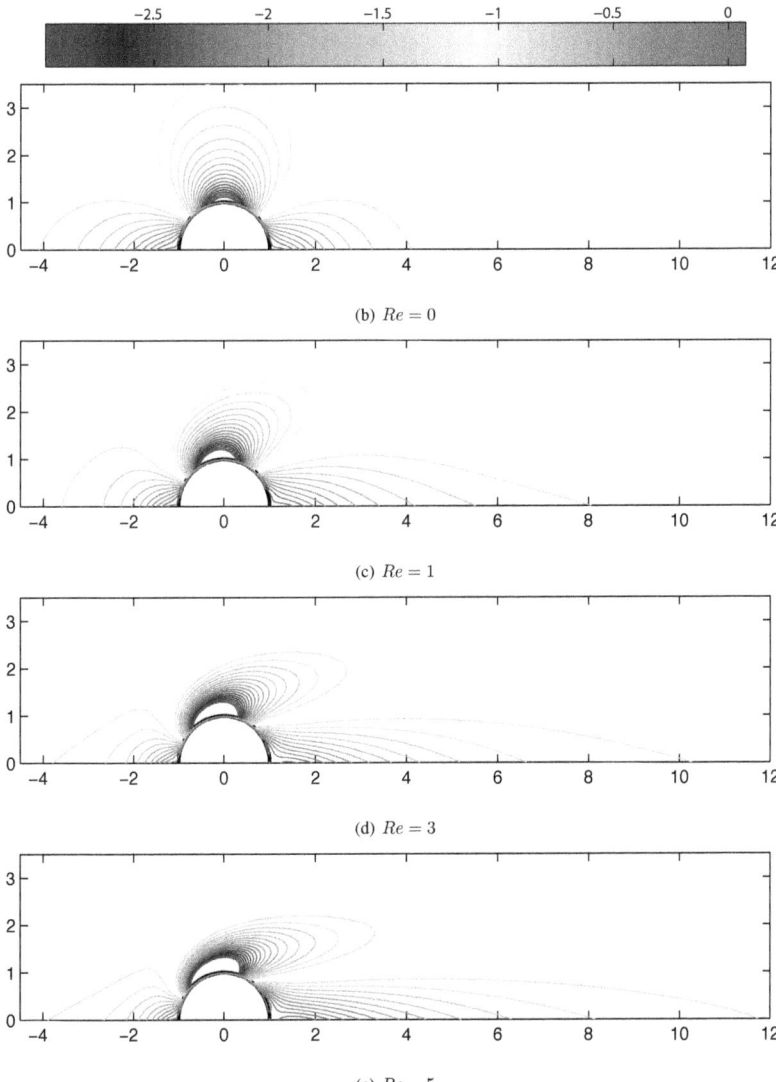

Figure 2.13: Lines of constant vorticity the steady flow in air over a semicircle with $h = \dfrac{\pi}{150}$ and a range of Reynolds number.

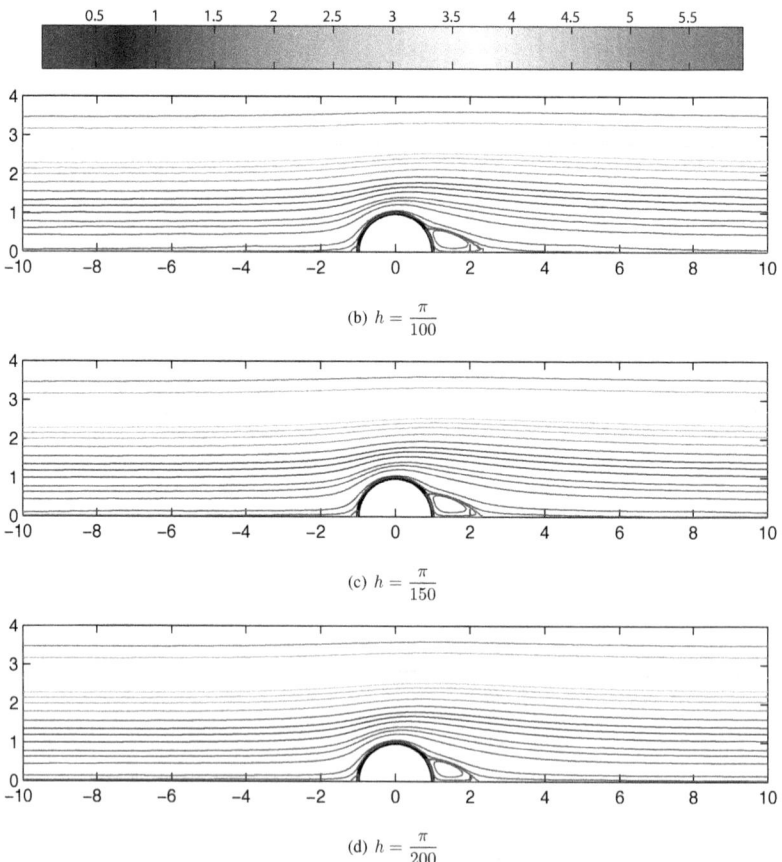

Figure 2.14: Grid refinement example for steady air flow over a semicircle: $\psi$ for $\widehat{Re} = 5$

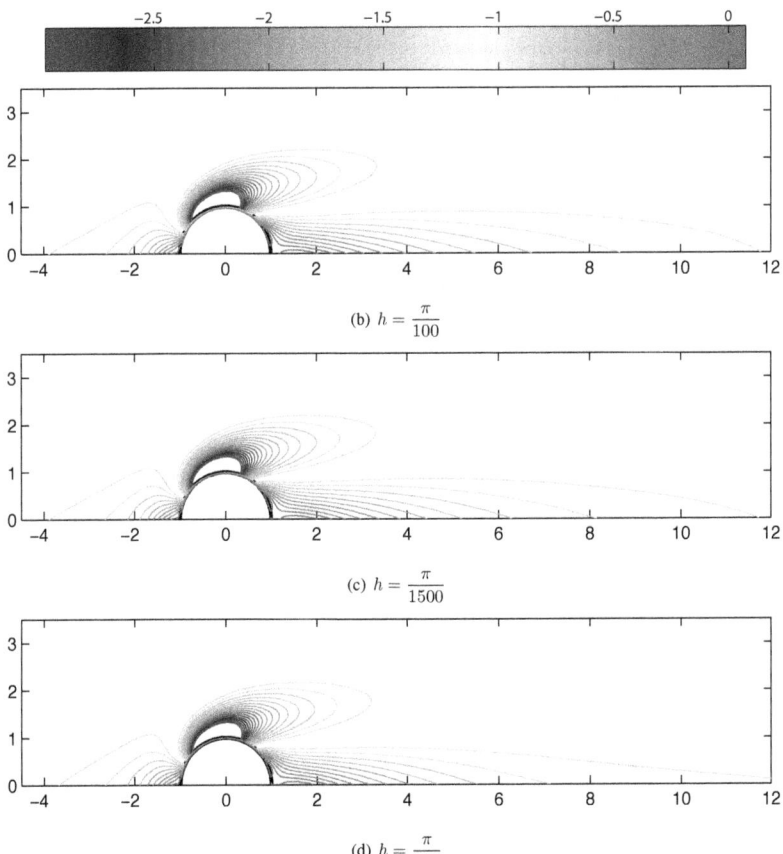

(b) $h = \dfrac{\pi}{100}$

(c) $h = \dfrac{\pi}{1500}$

(d) $h = \dfrac{\pi}{200}$

Figure 2.15: Grid refinement example for steady air flow over a semicircle: $\zeta$ for $\widehat{\mathrm{Re}} = 5$

vorticity which diffuses through the fluid. This was also found to occur in the later work done on the fluid within the droplet but since Rayleigh layers are not the main concern of this study, we omit any detailed analysis and refer the reader to Telionis [70], Stewartson [65] or Rayleigh [47]. Instead, to avoid this problem and the effect it has on the convergence of the time marching system, we chose instead to introduce a slowly started shear flow along $r_\infty$. This is done by replacing (2.2.7) with

$$\psi = \frac{1}{2}\frac{e^t}{1+e^t}r^2\sin^2\theta, \quad \zeta = -\frac{e^t}{1+e^t}, \quad \text{on } r = r_\infty. \tag{2.2.12}$$

and starting the system at $t = -5$. Aside from this feature, an identical process to the time march procedure detailed in the driven cavity is used to solve (2.2.8), (2.2.9) subject to the boundary conditions (2.2.5), (2.2.6) and (2.2.7).

Figure (2.16) shows the results for the time marched streamfunction for $t = 0.5, 1, 2, 4, 8$. The numerics are calculated using $\widehat{\text{Re}} = 3$ on a grid with spacing $h = \pi/150$, giving $151$ points in $\theta$ and $669$ in $r$. Contours are plotted for the values of $\psi$ listed in (2.2.10) so that the colour bar in figure (2.12a) also applies here. Our results suggest that by $t = 0.5$, a small recirculation zone downstream to the semicircle has already begun to form, increasing in size with time. The pinching of the streamlines is more severe for smaller values of $t$ signifying a larger streamfunction gradient above the obstacle. As time increases, we see that figure (2.16e) and the corresponding steady state result (2.12d) match well, suggesting that the flow solution becomes virtually steady between $t = 4$ and $t = 8$.

A similar trend may be seen for the accompanying vorticity solution in figure (2.16) where the colour bar in figure (2.13a) applies: by $t = 8$, a comparison between the time marched solution and the figure showing the corresponding steady state (2.13d) show similar trends in both. We conclude that the unsteady solution does indeed converge in time to the steady state solution detailed in the previous section.

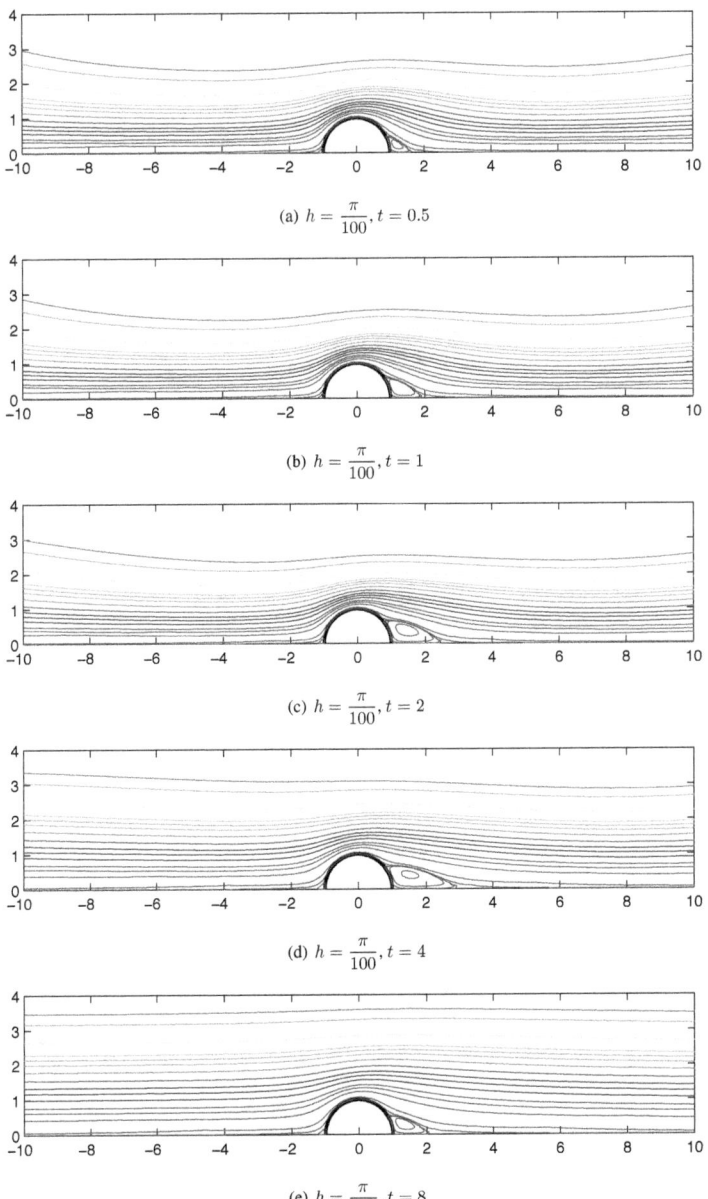

(a) $h = \dfrac{\pi}{100}, t = 0.5$

(b) $h = \dfrac{\pi}{100}, t = 1$

(c) $h = \dfrac{\pi}{100}, t = 2$

(d) $h = \dfrac{\pi}{100}, t = 4$

(e) $h = \dfrac{\pi}{100}, t = 8$

Figure 2.16: Unsteady results for $\psi$ for flow in air over a semicircle and $\widehat{\mathrm{Re}} = 3$

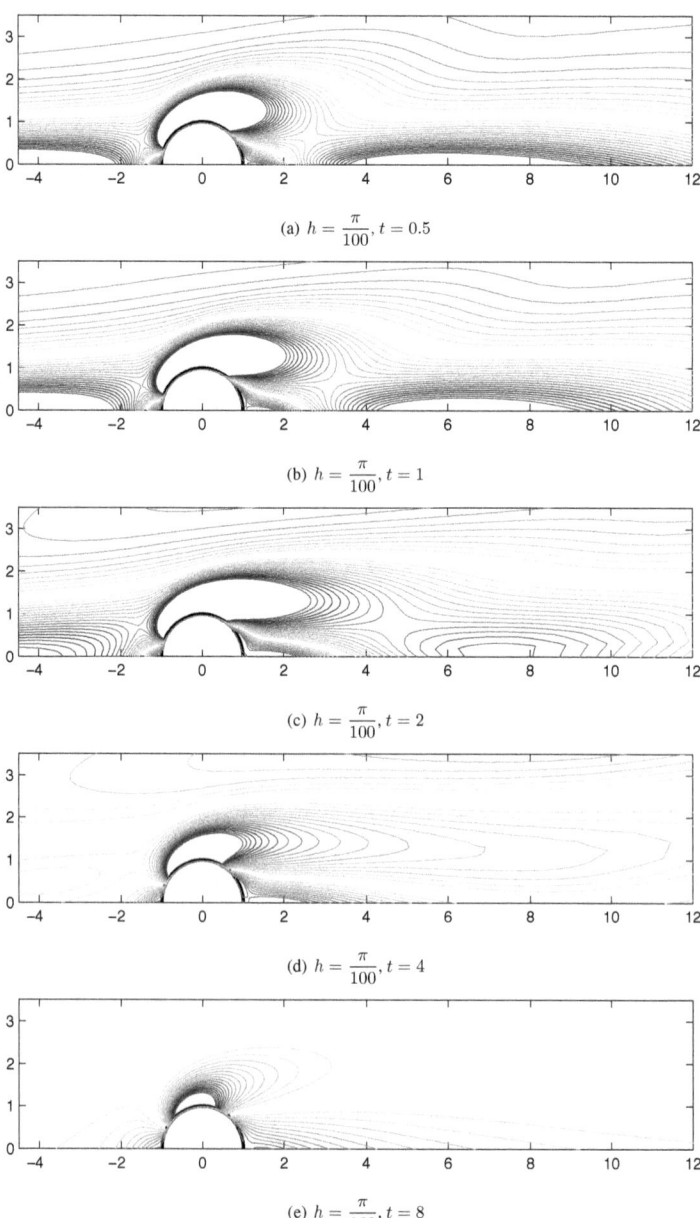

(a) $h = \dfrac{\pi}{100}, t = 0.5$

(b) $h = \dfrac{\pi}{100}, t = 1$

(c) $h = \dfrac{\pi}{100}, t = 2$

(d) $h = \dfrac{\pi}{100}, t = 4$

(e) $h = \dfrac{\pi}{100}, t = 8$

Figure 2.17: Unsteady results for $\zeta$ for flow in air over a semicircle and $\widehat{\mathrm{Re}} = 3$

With apparently sensible results for the time marching algorithm our next step is to extract

the pressure and vorticity along the wall of the obstacle, as outlined in the method description

given in Chapter 1. The vorticity along $r = 1$ forms part of our current solution, but the pres-

sure must be determined at each time step from the polar form of the Navier-Stokes equations.

Along the wall of the semicircle a no slip condition applies, all velocity terms are zero and the

Navier-Stokes in pressure vorticity form become the Cauchy-Riemann-like equations:

$$\frac{\partial p}{\partial r} = -\frac{1}{\widehat{Re} r}\frac{\partial \zeta}{\partial \theta} \qquad \frac{\partial p}{\partial \theta} = \frac{r}{\widehat{Re}}\frac{\partial \zeta}{\partial r}. \qquad (2.2.13)$$

The pressure may be found from the second of these:

$$p(r = 1, \theta) = \int_0^\theta \left.\frac{\partial \zeta(r, s)}{\partial r}\right|_{r=1} ds, \qquad (2.2.14)$$

where $p(1, 0) = 0$ is taken as a reference point. To apply this in a discrete form, we use

a Taylor's expansion of $\zeta$ at the two points interior to the boundary $((1 + h, \theta)$ labelled by

subscript $i$ and $(1 + 2h, \theta)$, labelled by subscript $j$). A linear combination of these allows us to

find the second order accurate $r$ derivative of $\zeta$ at $r = 1$ without using a shadow point:

$$\left.\frac{\partial \zeta}{\partial r}\right|_{r=1} = -\frac{(\zeta_j - 4\zeta_i + 3\zeta_b)}{2h}, \qquad (2.2.15)$$

here subscript $b$ is used to denote the line $(1, \theta)$ as before. The pressure may then be determined

by applying the trapezium rule to the result.

Figure (2.18a) shows a plot of the the time dependent interface pressure. For negative $t$

the pressure is relatively small, slowly increasing to a maximum at around $t = 0$. The pressure

takes a higher value at the leading edge of the droplet, $\theta = \pi$, throughout the simulation which

is as one would expect. Beyond this point, we see an area of negative pressure forming behind

the obstacle which matches well with the recirculation areas already seen in figure (2.16). As $t$

increases further the pressure solution indicates $\frac{\partial p}{\partial t} \to 0$ and the curve begins to flatten as the

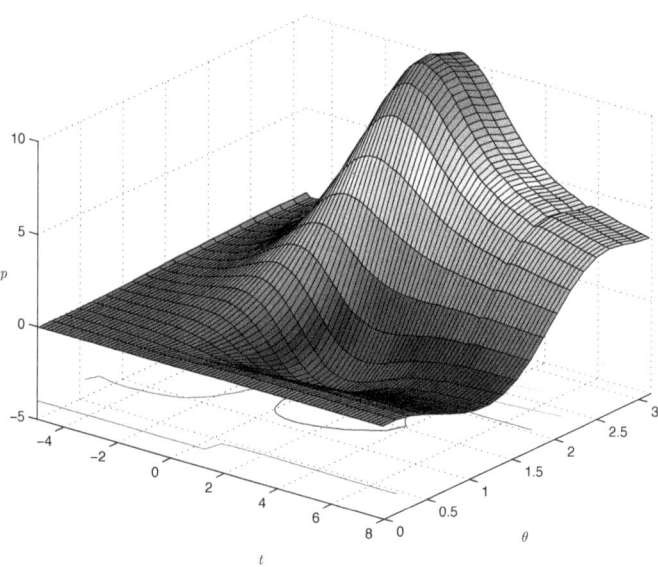

(a) The pressure along the wall of the obstacle as time increases

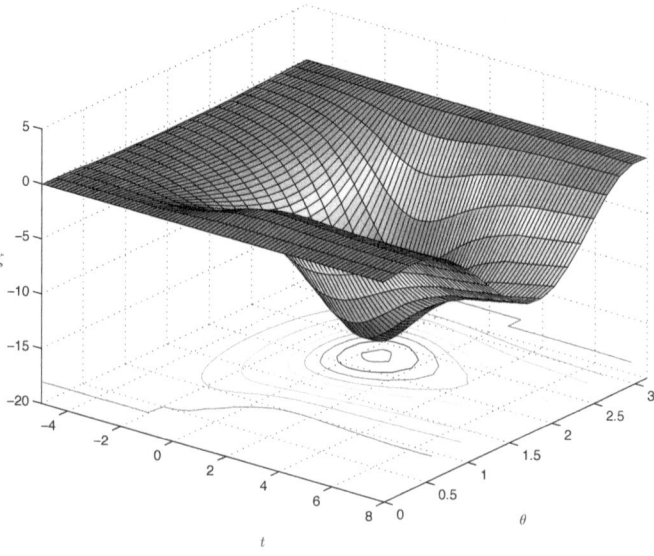

(b) The vorticity along the wall of the obstacle as time increases

Figure 2.18: The pressure and vorticity along interface for the flow in air over a semicircular obstacle.

time dependent form moves toward the steady state solution.

The corresponding image for vorticity is given in figure (2.18b). The general trends are similar: $\zeta$ takes small values at the intial time steps and the magnitude of $\zeta$ is at a maximum around $t = 0$. Our definition of vorticity in (2.2.3) fixes a clockwise rotation as $\zeta$ negative, thus the fluid particles along the top of the semicircle are rotating in the same direction as the external air flow. We notice an area of positive vorticity forming at the trailing edge of the semi circle, $\theta = 0$ which ties in with the recirculation discussed. Finally, the image begins to flatten along the time axis as $t \to 8$ and the solution to the unsteady problem becomes virtually steady.

With these results, step one of the method given in Chapter 1 is complete. We shall return to these solutions for $p$ and $\zeta$ when we attempt to tackle the interacting problem in Chapter 4, but first we concern ourselves with step two of the method: the flow within the semicircular water droplet.

## Chapter 3

# Flow in water

We now have an algorithm which allows us to determine the pressure and vorticity along the curved wall of a surface mounted semicircle at any time and so we switch our focus in this section to the flow within the droplet, corresponding to step two in the method outlined in Chapter 1.

As discussed there, a key concern of this study is the understanding of the behaviour of the droplet at larger times, particulary the identification of a later temporal stage when, to satisfy the kinematic condition, the relationship between the air and water becomes nonlinear. The time at which this change in dynamics occurs is dictated by the growth of the velocities within the droplet, as will become clear later in this chapter. If the velocities become steady the new time scale will come into operation as $t \to 1/\epsilon$, where $\epsilon$ is our small density ratio. However, if linear velocity growth with time is found, this later stage is as $t \to 1/\epsilon^{1/2}$.

It proves difficult to confirm directly the behavior of the velocities at larger times. To better understand this mechanism it is helpful to include the analysis of two additional canonical problems. First, we study an entirely free surface droplet, the circular seed; and second, a Cartesian version of our aim, the square droplet. These model problems offer insight and validation for the semicircular case, which we tackle in Section 3.3.

As a further simplification we temporarily abandon the accurate but numerical pressure and vorticity results from the previous chapter and instead prescribe idealised pressure and vorticity interface conditions along the free surface of the droplet. If the fluid is initially at rest, we find that doing so causes the undesired effect of Rayleigh layers. To avoid this, we allow for a gentle introduction of the $p$ and $\zeta$ interface conditions by including a slowly increasing function of time which tends to unity, similar to that seen in (2.2.12).

Throughout this chapter, we work with the dimensionless Stokes equations, found from applying the small ratios scaling $(\hat{\mathbf{u}}_W, \hat{p}_W) = \epsilon(\mathbf{u}_W, p_W)$ to (1.2.10), which were given in Chapter 1 but we repeat here:

$$\frac{\partial u}{\partial t} = -\frac{\partial p}{\partial x} - \mathrm{Re}^{-1}\frac{\partial \zeta}{\partial y}, \tag{3.0.1}$$

$$\frac{\partial v}{\partial t} = -\frac{\partial p}{\partial y} + \mathrm{Re}^{-1}\frac{\partial \zeta}{\partial x}, \tag{3.0.2}$$

$$\frac{\partial u}{\partial x} + \frac{\partial v}{\partial y} = 0, \tag{3.0.3}$$

where subscript $W$ is dropped for ease of notation. With the vorticity defined as

$$\zeta = \frac{\partial v}{\partial x} - \frac{\partial u}{\partial y}, \tag{3.0.4}$$

it is relatively simple to manipulate equations (3.0.1), (3.0.2) into a vorticity-pressure system:

$$\nabla^2 p = 0, \tag{3.0.5}$$

$$\frac{\partial \zeta}{\partial t} = \mathrm{Re}^{-1}\nabla^2\zeta. \tag{3.0.6}$$

Further, along a wall where a no-slip condition applies and the velocities $u, v$ are identically zero equations (3.0.1) and (3.0.2) become the Cauchy-Riemann equations for pressure and vorticity:

$$\frac{\partial p}{\partial x} = -\mathrm{Re}^{-1}\frac{\partial \zeta}{\partial y}, \tag{3.0.7}$$

$$\frac{\partial p}{\partial y} = \mathrm{Re}^{-1}\frac{\partial \zeta}{\partial x}. \tag{3.0.8}$$

In the circular seed model and the semicircular droplet, the equations become simplest in polar form. In such a case, where the Laplacian is:

$$\nabla^2 p = \frac{1}{r}\frac{\partial}{\partial r}\left(r\frac{\partial p}{\partial r}\right) + \frac{1}{r^2}\frac{\partial^2 p}{\partial \theta^2}, \tag{3.0.9}$$

equations (3.0.5) and (3.0.6) remain unchanged, and with vorticity defined as

$$\zeta = \frac{1}{r}\frac{\partial}{\partial r}(ru_\theta) - \frac{1}{r}\frac{\partial}{\partial \theta}(u_r), \tag{3.0.10}$$

the field equations (3.0.1), (3.0.2) become

$$\frac{\partial u_r}{\partial t} = -\frac{\partial p}{\partial r} - \frac{\mathrm{Re}^{-1}}{r}\frac{\partial \zeta}{\partial \theta}, \tag{3.0.11}$$

$$\frac{\partial u_\theta}{\partial t} = -\frac{1}{r}\frac{\partial p}{\partial \theta} + \mathrm{Re}^{-1}\frac{\partial \zeta}{\partial r}. \tag{3.0.12}$$

The continuity equation is

$$\frac{1}{r}\frac{\partial}{\partial r}(ru_r) + \frac{1}{r}\frac{\partial u_\theta}{\partial \theta} = 0, \tag{3.0.13}$$

and a no slip condition along a wall again yields the Cauchy-Riemann equations:

$$\frac{\partial p}{\partial r} = -\frac{\mathrm{Re}^{-1}}{r}\frac{\partial \zeta}{\partial \theta}, \tag{3.0.14}$$

$$\frac{1}{r}\frac{\partial p}{\partial \theta} = \mathrm{Re}^{-1}\frac{\partial \zeta}{\partial r}. \tag{3.0.15}$$

## 3.1 The circular seed

Using the system outlined above, the first model problem we consider is of the flow within a free surface circular droplet with a solid center, subject to a unidirectional external flow field. The solid 'seed' and the droplet form two concentric circles with the given solid-liquid interface lying at $r = a$ as shown in figure (3.1). The droplet has symmetry about the line $y = 0$, and as such, we need only model fluid contained in the half plane $y > 0$. The solid center simplifies the problem further, allowing us to use a no slip condition at $r = a$, which reduces the system to a one dimensional problem. Of course, as $a \to 0$, one would expect our solution to approach that of an entirely free surface droplet.

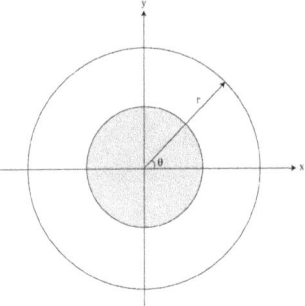

Figure 3.1: A diagramatical representation of the circular seed problem.

By choosing free surface conditions

$$p(1,\theta) = \frac{e^t}{1+e^t} \sin\theta, \tag{3.1.1}$$

$$\zeta(1,\theta) = \frac{e^t}{1+e^t} \sin\theta, \tag{3.1.2}$$

a full description of the flow is given by the coupled system (3.0.5), (3.0.6) subject to the boundary conditions (3.1.1) and (3.1.2) along $r = 1$, and the Cauchy-Riemann equations (equivalent to a no slip condition) (3.0.14) and (3.0.15) along $r = a$.

This problem may be solved using a trigonometric series for the pressure and vorticity. It turns out, to satisfy the momentum equations within the droplet and the no slip conditions along the liquid-solid boundary we need only the first term in the series. The uniqueness of this solution is discussed later. We also know what form the coefficients for pressure must take to ensure it satisfies Laplace's equation, so that pressure and vorticity may be described by

$$p = \left(Ar + \frac{\tilde{A}}{r}\right)\sin\theta + \left(Br + \frac{\tilde{B}}{r}\right)\cos\theta, \tag{3.1.3}$$

$$\zeta = C(r,t)\sin\theta + D(r,t)\cos\theta, \tag{3.1.4}$$

for unknown constants $A, \tilde{A}, B, \tilde{B}$ and functions $C(r,t)$, $D(r,t)$. Rewriting $g(t) = \dfrac{e^t}{1+e^t}$, with the pressure and vorticity in this form, the choice of interface condition immediately suggests gives $B + \tilde{B} = 0$, and $A + \tilde{A} = g(t)$.

Applying (3.1.3), (3.1.4) to the field equations and boundary conditions listed above leads to two decoupled evolution equations for $C$ and $D$.

For $C(r,t)$:

$$\mathrm{Re}^{-1}\left[\frac{\partial^2 C}{\partial r^2} + \frac{1}{r}\frac{\partial C}{\partial r} - \frac{1}{r^2}C\right] = \frac{\partial C}{\partial t}, \tag{3.1.5}$$

$$C = g(t) \quad \text{on } r = 1, \text{ which} \to 1 \text{ as } t \to \infty, \tag{3.1.6}$$

$$\left.\begin{aligned}
\frac{\partial C}{\partial r} &= \mathrm{Re}B\left(\frac{1}{r^2} - 1\right) \\[2ex]
C &= -r\mathrm{Re}B\left(1 + \frac{1}{r^2}\right)
\end{aligned}\right\} \quad \text{on } \ r = a. \tag{3.1.7}$$

For $D(r,t)$:

$$\mathrm{Re}^{-1}\left[\frac{\partial^2 D}{\partial r^2} + \frac{1}{r}\frac{\partial D}{\partial r} - \frac{1}{r^2}D\right] = \frac{\partial D}{\partial t}, \tag{3.1.8}$$

$$D = 0 \quad \text{on } r = 1, \tag{3.1.9}$$

$$\left.\begin{aligned}
\frac{\partial D}{\partial r} &= \mathrm{Re}\left(A + \frac{\tilde{A}}{r^2}\right) \\[2ex]
D &= r\mathrm{Re}\left(A - \frac{\tilde{A}}{r^2}\right)
\end{aligned}\right\} \quad \text{on } \ r = a. \tag{3.1.10}$$

The constants $A, \tilde{A}$ and $B, \tilde{B}$ may be eliminated from the conditions along $r = a$, so that rather than (3.1.7), $C(a,t)$ can be found from:

$$\frac{1}{C}\frac{\partial C}{\partial r}\bigg|_{r=a} = \frac{(a^2 - 1)}{a(a^2 + 1)}, \tag{3.1.11}$$

and, rather than (3.1.10), $D(a,t)$ may be found from

$$(1 - a^2)D + (a^2 + 1)a\frac{\partial D}{\partial r}\bigg|_{r=a} = 2ag(t)\mathrm{Re}. \tag{3.1.12}$$

### 3.1.1 The time dependent problem

The two sets of equations ((3.1.5), (3.1.6), (3.1.11) for $C$ and (3.1.8), (3.1.9), (3.1.12) for $D$) may be solved separately by using a semi-implicit finite difference scheme, second order accurate in space and first order accurate in time, similar to that seen in the lid driven cavity problem of the previous chapter and precise details of which we do not feel it necessary to include here. To deal with the $r$ derivative at $r = a$ in (3.1.11) and (3.1.12) we use a Taylor's expansion of the two grid points immediately interior to the boundary and arrive at an expression identical in form to that seen in (2.2.15).

Once a numerical solution to $C$ and $D$ has been determined at a given time step, pressure and vorticity may be found directly, and substituted back into the momentum equation (3.0.11) to find $u_r$ along the free surface. The kinematic condition provides the new interface shape as discussed later. Finally, the constants $A, \tilde{A}, B, \tilde{B}$ are given by,

$$A = \frac{aD(a,t) + g(t)}{\mathrm{Re}(1 + a^2)}, \tag{3.1.13}$$

$$B = -\tilde{B} = \frac{-C(a,t)a}{a^2 + 1}, \tag{3.1.14}$$

$$\tilde{A} = \frac{g(t)a^2 - a\mathrm{Re}^{-1}D(a,t)}{1 + a^2}. \tag{3.1.15}$$

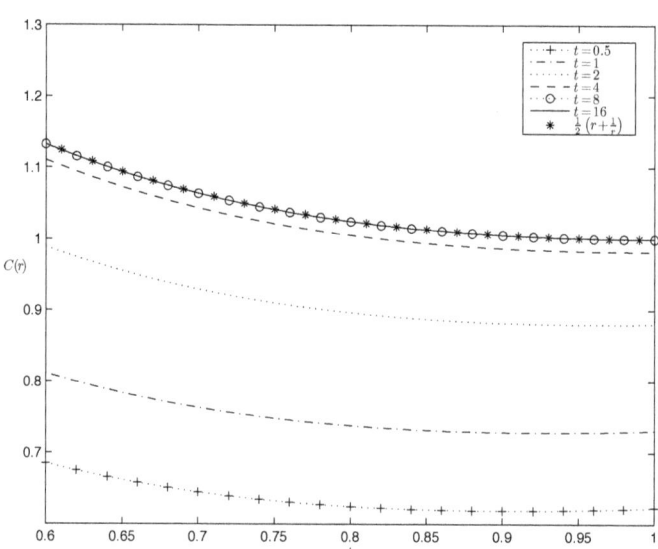

Figure 3.2: $C(r,t)$ against $r$ at different times, and $\dfrac{1}{2}\left(r + \dfrac{1}{r}\right)$.

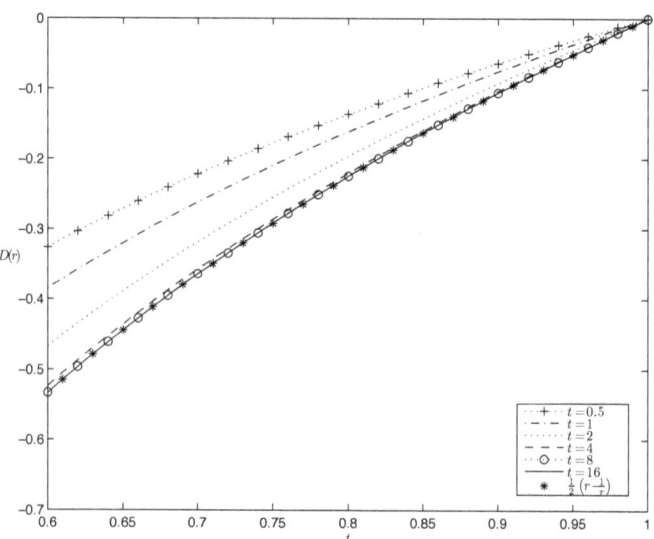

Figure 3.3: $D(r,t)$ against $r$ at different times, and $\dfrac{1}{2}\left(r - \dfrac{1}{r}\right)$.

Plots are shown in figure (3.2) for $C$ against $r$ for increasing time. Likewise for $D$ in figure (3.3). In all of our numerical work the constants $(a, \mathrm{Re}, \delta r, \delta\theta) = (0.6, 1, 0.005, 0.005)$ were used, to give 81 grid points in $r$ and 629 points in $\theta$. We can see from the graphs that $C$ and $D$ both converge to a steady state in effect sometime between $t = 4$ and $t = 8$ and remain steady at least until $t = 16$. The exact solutions of the steady state are also plotted as solid lines with black stars, and these are discussed later.

Plots are included of the pressure and vorticity induced along the solid boundary $r = a$ for increasing time in figures (3.4) and (3.5) respectively. Along this line the boundary conditions dictate that the pressure and vorticity must satisfy the Cauchy-Riemann equations. It appears from the graph that the vorticity is simply the pressure reflected in the line $\theta = \pi/2$, so that $\zeta(r, \theta, t) = p(r, \pi - \theta, t)$. Both graphs also become effectively steady sometime between $t = 4$ and $t = 8$ and remain steady for larger times. In these images, $\theta = 0$ refers to the positive $x$ axis and thus in a sense the trailing edge of the droplet.

Figure (3.6) shows the radial velocity produced along the free surface of the droplet. Likewise in figure (3.7) for the azimuthal velocity. Crucially we notice that in addition to the pressure and vorticity becoming steady, the numerical results suggest that $\mathbf{u}$ becomes steady with time also. This has a direct impact on the time scale of the later temporal stage, signifying a nonlinear interaction between air and water. We discuss this in a moment, but first briefly present an argument for the uniqueness of this solution.

### 3.1.2 Uniqueness of the solution

Consider the other terms in an infinite series solution to the Laplacian,

$$\hat{p} = \sum_{n=2}^{\infty} \left( A_n r^n - \frac{\tilde{A}_n}{r^n} \right) \sin n\theta + \left( B_n r^n - \frac{\tilde{B}_n}{r^n} \right) \cos n\theta. \tag{3.1.16}$$

Likewise, the other terms in an infinite series of $\zeta$ are

$$\hat{\zeta} = \sum_{n=2}^{\infty} C_n(r,t) \sin n\theta + D_n(r,t) \cos n\theta. \tag{3.1.17}$$

Working through the equations and boundary conditions with these extra terms, if all these coefficients can be shown to be zero, our solution found from the first term in the series must be unique.

Applying the boundary conditions at $r = 1$ and $r = a$ gives $A_n = -\tilde{A}_n$, $B_n = -\tilde{B}_n$, and the following relationship for $A_n$ and $D_n$

$$\left.\frac{\partial D_n}{\partial r}\right|_{r=a} = nA_n \left(a^{n-1} - \frac{1}{a^{n+1}}\right), \tag{3.1.18}$$

$$\frac{D_n}{r} = aA_n \left(a^{n-1} + \frac{1}{a^{n+1}}\right). \tag{3.1.19}$$

Aside from an extra minus sign on the left hand side in each equation, these take the same form as the $C_n, B_n$ set.

In the same way as before, $A_n$ can be eliminated, giving a expression for $D_n$ alone.

$$\left.\frac{1}{D_n}\frac{\partial D_n}{\partial r}\right|_{r=a} = \frac{n\left(a^{2n} - 1\right)}{a\left(a^{2n} + 1\right)}. \tag{3.1.20}$$

The right hand side of (3.1.20) is time independent, and the left contains only a derivative in $r$. For this to be true, $D_n$ must take the form $D_n = T_n(t)R_n(r)$. We may apply this in the decoupled evolution equation for $D_n$,

$$\mathrm{Re}^{-1}\left[\frac{\partial^2 D_n}{\partial r^2} + \frac{1}{r}\frac{\partial D_n}{\partial r} - \frac{n^2}{r^2}D_n\right] = \frac{\partial D_n}{\partial t}, \tag{3.1.21}$$

to obtain a solution for $T_n(t)$ using standard techniques,

$$T_n(t) = \tau_n \exp[-\lambda_n^2 t], \tag{3.1.22}$$

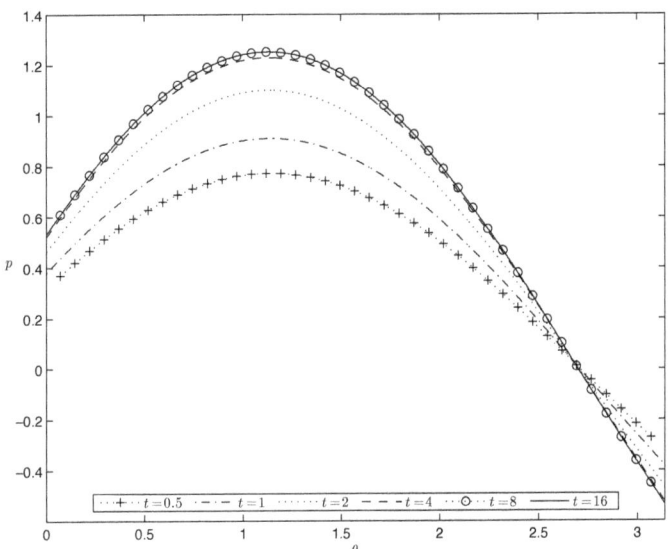

Figure 3.4: The pressure $p$ against $\theta$ along the solid boundary, $r = a$.

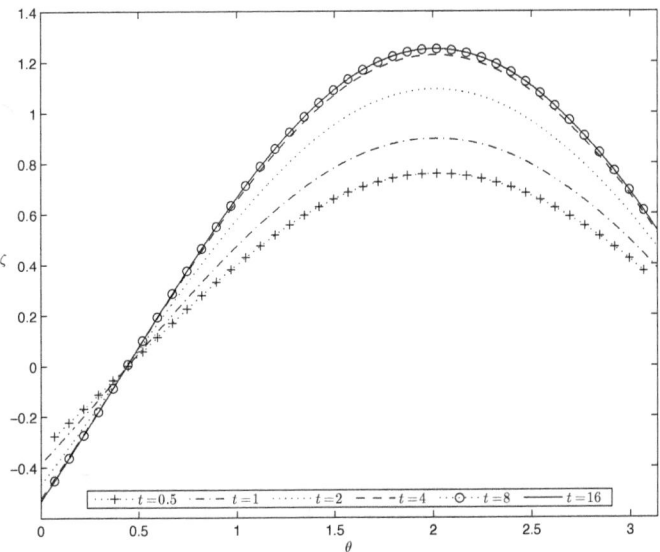

Figure 3.5: The vorticity $\zeta$ against $\theta$ along the solid boundary, $r = a$.

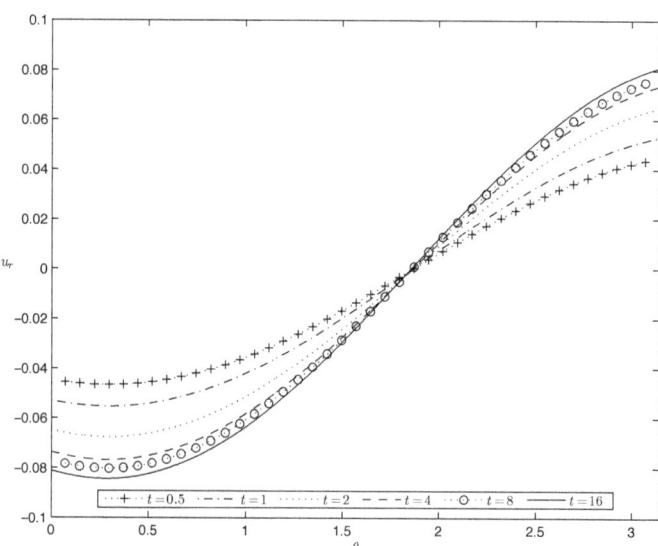

Figure 3.6: Radial velocity $u_r$ against $\theta$ along the two fluid interface $r = 1$.

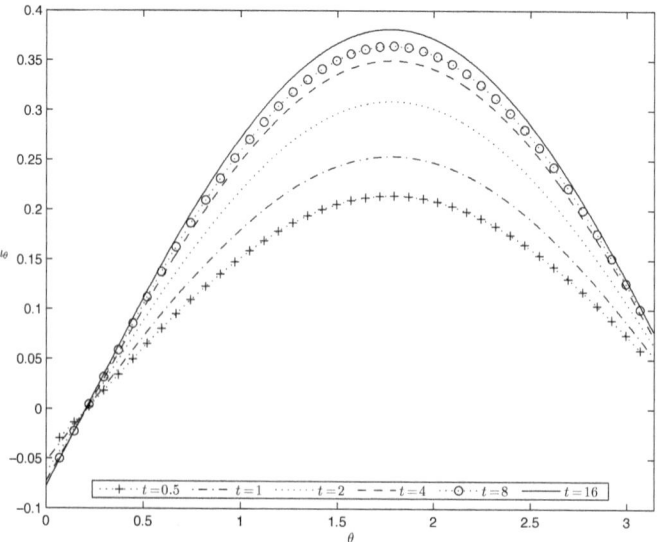

Figure 3.7: Azimuthal velocity $u_\theta$ against $\theta$ along the two fluid interface $r = 1$.

for real constants $\lambda_n$ and $\tau_n$.

If the external flow field is impulsively started, the vorticity would be initially zero within the droplet. For this to be true, $D_n = 0$ at $t = 0$ making $\tau_n = 0$. From there, $T(t)$ is zero for all time, as is $D_n$ and hence $A_n, \tilde{A}_n = 0$ from (3.1.19).

The same argument holds for $C_n, B_n$, so that they too must be zero for all time. All coefficients for $n \geq 2$ are zero and remain zero, which makes our first term only solution unique.

### 3.1.3 Shape analysis and the later temporal stage

Initially, the interface between the water and air is described by the circle $f(\theta, t) = 1$. It seems reasonable to suggest that the droplet deformation will be dependent on the small density ratios parameter $\epsilon$. So, we expect the shape to expand as $f = f_0 + \epsilon f_1 + \epsilon^2 h_2 + \cdots$ in the early stage.

To derive the kinematic condition in polar form, we take the time derivative of $r - f(\theta, t) = 0$. Then,

$$\frac{\partial r}{\partial t} - \frac{\partial f}{\partial \theta}\frac{\partial \theta}{\partial t} - \frac{\partial f}{\partial t} = 0, \tag{3.1.23}$$

which is equivalent to

$$\hat{u}_r = \hat{u}_\theta \frac{\partial f}{\partial \theta} + \frac{\partial f}{\partial t}. \tag{3.1.24}$$

The scalings for velocity lead to an $O(\epsilon)$ balance of the kinematic condition of

$$u_r = u_\theta \frac{\partial f_0}{\partial \theta} + \frac{\partial f_1}{\partial t}, \tag{3.1.25}$$

suggesting that the leading order shape effect $f_0$ is time independent. Further, since $f_0 = 1$, the kinematic condition here just becomes

$$u_r = \frac{\partial f_1}{\partial t}, \tag{3.1.26}$$

which yields the new droplet shape from a first order accurate finite difference procedure

$$f_1^k = \delta t u_r^{k-1} + f_1^{k-1}. \tag{3.1.27}$$

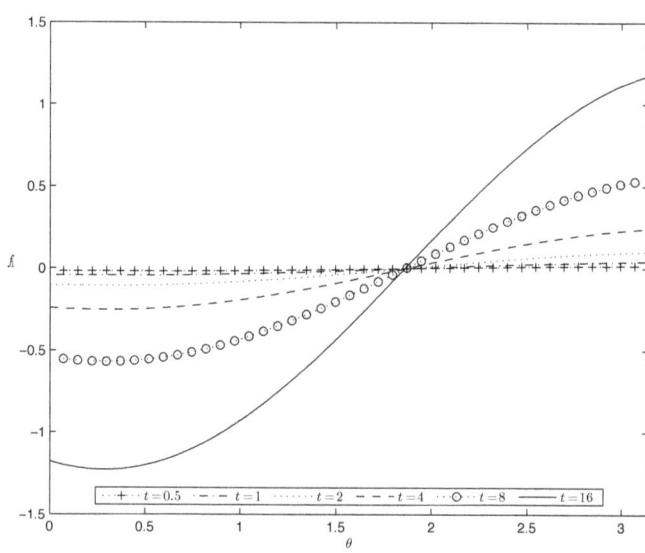

Figure 3.8: The shape effect, $f_1$ of the circular seed problem at different times.

The numerical results appear to suggest that all major variables (apart from $f_1$) approach a steady state at large times and remain steady. Our argument in Section 3.1.2 indicates that this solution is unique. For a steady $u_r$, (3.1.26) implies that $f_1$ will grow linearly with time which can be seen to match our numerical results in figure (3.8), where $f_1$ is plotted at $t = 0.5, 1, 2, 4, 8, 16$.

This model will hold until $t$ approaches $O(1/\epsilon)$, and $f_1$ becomes comparable with the leading order term in the interface shape, $f_0$, and the kinematic boundary condition becomes fully nonlinear. This corresponds to our second temporal stage. A rescaling of $t = T/\epsilon$ would allow us to explore the dynamics of this stage, we do so briefly in Section 3.3 for the semicircular droplet and numerically in Part 2 for the triple deck droplet, but further analysis is not provided for the current model.

### 3.1.4 The steady state problem

If the vorticity does reach a steady state (and satisfies Laplace's equation as opposed to the heat equation) then $C$ and $D$ become $C(r)$ and $D(r)$. The equations (3.1.5), (3.1.6), (3.1.11), and (3.1.8), (3.1.9), (3.1.12) still hold, but with $g(t) = 1$. These yield an exact solution for $C$ and $D$ of

$$C(r) = \left(\frac{r}{2} + \frac{1}{2r}\right), \qquad\qquad D(r) = \left(\frac{r}{2} - \frac{1}{2r}\right). \qquad (3.1.28)$$

Then $A$, $\tilde{A}$, $B$ and $\tilde{B}$ may be uniquely determined from this solution, which gives steady state results for the pressure and vorticity as

$$p = \frac{1}{2}\left(r + \frac{1}{r}\right)\sin\theta - \frac{1}{2}\left(r - \frac{1}{r}\right)\cos\theta, \qquad (3.1.29)$$

$$\zeta = \frac{1}{2}\left(r + \frac{1}{r}\right)\sin\theta + \frac{1}{2}\left(r - \frac{1}{r}\right)\cos\theta. \qquad (3.1.30)$$

The steady state solutions of $C$ and $D$ in (3.1.28) are plotted in figures (3.2) and (3.3) alongside their numerically calculated time-marched results and shown as solid lines with black stars. The numerical results closely approach the steady state exact solution as $t$ increases, which suggests that a steady state is reached and that our numerical procedure is accurate. It is worth noting also that with this solution $\zeta(r, \theta) = p(r, \pi - \theta)$, which agrees with the results for the pressure and vorticity shown in figures (3.4) and (3.5).

The velocity will remain steady as long as $\partial_t \mathbf{u} = 0$ of course. For this to be true, the momentum equations (3.0.11) and (3.0.12) dictate that the pressure and vorticity must satisfy the Cauchy-Riemann equations everywhere. Thinking in Cartesian terms,

$$\mathrm{Re}^{-1}\frac{\partial\zeta}{\partial x} = \frac{\partial p}{\partial y}, \qquad -\mathrm{Re}^{-1}\frac{\partial\zeta}{\partial y} = \frac{\partial p}{\partial x}. \qquad (3.1.31)$$

Two functions which satisfy the Cauchy-Riemann equations must both be harmonic (satisfy Laplace's equation). Further, there must exist an analytic function $F(z)$, of $z = x + iy$, where $F(z) = \mathrm{Re}^{-1}\zeta + ip$, as outlined in Priestley [45]

To find such a function in the circular seed problem, we begin by rewriting (3.1.29) and

(3.1.30) in Cartesian coordinates:

$$p = \frac{1}{2}(y - x) + \frac{1}{2}\frac{x + y}{x^2 + y^2}, \tag{3.1.32}$$

$$\zeta = \frac{1}{2}(x + y) + \frac{1}{2}\frac{y - x}{x^2 + y^2}. \tag{3.1.33}$$

Taking $\mathrm{Re} = 1$, we find an analytic function $F(z) = \zeta + ip$ as follows:

$$F(z) = \frac{1}{2}(x + y) + \frac{1}{2}\frac{y - x}{x^2 + y^2} + i\frac{1}{2}(y - x) + i\frac{1}{2}\frac{x + y}{x^2 + y^2}, \tag{3.1.34}$$

$$= \frac{1}{2}(x + iy) + \frac{1}{2}(y - ix) + \frac{1}{2}\frac{y + ix}{x^2 + y^2} - \frac{1}{2}\frac{x + iy}{x^2 + y^2}, \tag{3.1.35}$$

$$= \frac{z}{2} - \frac{iz}{2} + \frac{iz^*}{2zz^*} - \frac{z^*}{2zz^*}, \tag{3.1.36}$$

$$= \frac{1}{2}(1 - i)\left(z - \frac{1}{z}\right). \tag{3.1.37}$$

So, once a steady state solution is reached, $p$ and $\zeta$ satisfy the Cauchy-Riemann equations ev-

erywhere and the system remains steady. Equally, the reverse is generally true for any problem

of this sort; to find an analytic function $F(z)$ independent of time implies that the velocities

remain steady, we see linear growth in $f_1$ and the later temporal stage comes into operation as

$t = O(1/\epsilon)$.

## 3.2   The square droplet

The circular seed offered us some insight into the dynamics of the flow within the water droplet.

We found that along a line where a no slip condition is present (and hence the Cauchy-Riemann

equations for $p$ and $\zeta$ hold) the pressure and vorticity form solutions which are a reflection of

one another about the vertical line through the midpoint of the geometry. Further, once the

flow solution reaches a steady state, so too do the velocities, implying backwards and forwards

that pressure and vorticity satisfy the Cauchy-Riemann equations everywhere and the shape of

the droplet changes linearly with time. These are potentially important results, but since the

circular seed case reduces to a one-dimensional problem we perhaps find their relevance in applying them to the semicircular droplet is limited. For that reason we now include a second

canonical problem: the square droplet. This may be thought of as a Cartesian version of our

semicircular droplet case, where the line $y = 0$ corresponds to $\theta = (0, \pi)$, $y = 1$ corresponds

to $r = 1$ and $x = (0, 1)$ are the points on the border of the semicircle near $r = 1, \theta = (0, \pi)$

respectively. The square droplet also has an identical geometry to the lid driven cavity outlined

earlier in this part of the thesis in Chapter 2 and a diagram of the setup may be seen in figure

(3.9). While physically, surface tension and gravitational effects limit the application of this

example to a real world problem (see Chapter 7), it does offer further insight into the dynamics

of the semicircular case, in particular into the behaviour of the velocities at large time as we

shall see.

We begin as we did in the circular seed, with the system of equations outlined at the beginning

of the chapter. Choosing a prescribed interface condition along $y = 1$:

$$p(x, 1, t) = \frac{e^t}{1 + e^t} \sin^2 \pi x, \tag{3.2.1}$$

$$\zeta(x, 1, t) = \frac{e^t}{1 + e^t} \sin^2 \pi x, \tag{3.2.2}$$

and ambient pressures and zero vorticity along $x = 0, 1$,

$$p = \zeta = 0 \text{ on } x = 0, 1, \tag{3.2.3}$$

a complete description of the problem is then given by the equations (3.0.5), (3.0.5) subject to

the boundary conditions (3.2.1), (3.2.2) on $y = 1$, (3.2.3) on $x = 0, 1$ and (3.0.8), (3.0.7) along

$y = 0$, equivalent to a no slip condition.

### 3.2.1 The steady state problem

In the same way as for the circular seed, we propose that the square droplet will converge in

time to a unique steady solution for pressure and vorticity. If this is the case, finding a steady

solution will allow us to later validate our time marching results. Further, if the steady state

solutions for $p$ and $\zeta$ may be written as an analytic function $F(z) = \text{Re}^{-1}\zeta + ip$, the circular

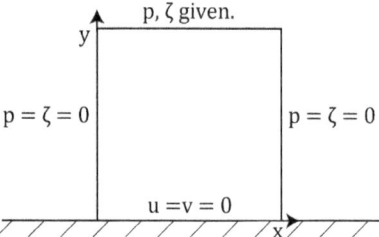

Figure 3.9: A diagramatical representation of the square droplet problem and its boundary conditions.

seed results suggest we to expect steady velocities and linear growth in the interface $f$. By assuming that the time dependent term in (3.0.6) tends to zero as time increases, the governing equations reduce to Laplace's equation for both pressure and vorticity, and we may approach a solution in two ways. The first is via a series solution and the second by using a numerical algorithm which is extended to form the basis of our time marching method in the next section. Comparisons between the results from the numerical algorithm and series solutions then will give us additional confidence in our time marching procedure of later.

**A series solution**

With the problem in the form above (Laplace's equation for pressure and vorticity and boundary conditions (3.0.8), (3.0.7), (3.2.3), (3.0.1) and (3.0.2)) we notice the similarities to the Fourier series solution of steady heat transfer in a square conducting plate. Indeed, by taking pressure or vorticity as analogous to temperature, we may build our own solution from two such Fourier problems, equivalent to first heating along $y = 0$, and second along $y = 1$, while keeping all other sides of the square at zero throughout the calculations. Figure (3.10) demonstrates this idea diagrammatically for the pressure: we solve for (a) and (b) separately, and sum together to achieve the full series solution.

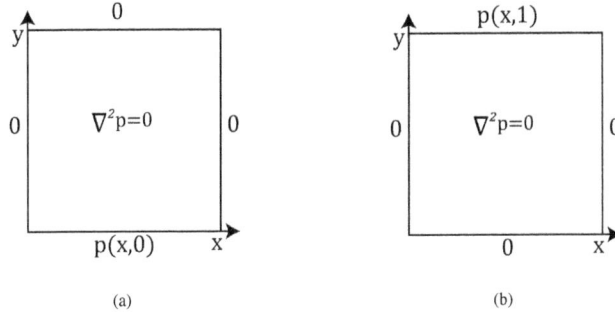

Figure 3.10: The series solution is built from the sum of two similar problems

In the case of pressure then, following the method outlined in Carslaw [11], we begin by considering a condition along $y = 0$ which may be written as a sine series:

$$p(x, 0) = \sum_n p_n \sin n\pi x, \qquad (3.2.4)$$

and fix $p$ as zero along $x = 0, 1$ and $y = 1$ as in figure (3.10a). With Laplace's equation (3.0.5) governing the pressure throughout the square, this setup has solution

$$p = \sum \frac{p_n \sinh\left[(1 - y)n\pi\right] \sin n\pi x}{\sinh n\pi}. \qquad (3.2.5)$$

Likewise, a series solution may be found from taking $p$ as zero along $x = 0, 1$ and $y = 0$ as in figure (3.10b), this time taking the prescribed pressure along $y = 1$ as of the form

$$p(x, 1) = \sum_n q_n \sin n\pi x. \qquad (3.2.6)$$

The series which satisfies Laplace's equation and these boundary conditions is simply

$$p = \sum_n \frac{q_n \sinh\left[n\pi y\right] \sin n\pi x}{\sinh n\pi}. \qquad (3.2.7)$$

Finally, the pressure for our case (equation (3.0.5) and boundary conditions (3.2.1), (3.2.3), (3.0.8) and (3.0.7)) is the sum of these two expressions (3.2.7) and (3.2.5):

$$p = \sum_n \frac{\left(p_n \sinh\left[n\pi(1 - y)\right] + q_n \sinh\left[n\pi y\right]\right) \sin n\pi x}{\sinh n\pi}. \qquad (3.2.8)$$

Further, since (3.2.6) implies

$$q_n = 2 \int_0^1 p(x, 1) \sin n\pi x \, dx, \tag{3.2.9}$$

$q_n$ may be determined by applying the steady state version of our choice of prescribed pressure (3.2.1) and integrating, giving the exact result

$$q_n = \frac{4\left((-1)^n - 1\right)}{\pi \left(n^3 - 4n\right)}. \tag{3.2.10}$$

An identical method may be used to find an expression for the vorticity:

$$\zeta = \sum_n \frac{\left(r_n \sinh\left[n\pi(1 - y)\right] + s_n \sinh n\pi y\right) \sin n\pi x}{\sinh n\pi}, \tag{3.2.11}$$

where

$$r_n = 2 \int_0^1 \zeta(x, 0) \sin n\pi x \, dx, \tag{3.2.12}$$

$$s_n = 2 \int_0^1 \zeta(x, 1) \sin n\pi \, dx, \tag{3.2.13}$$

and our choice of prescribed vorticity of earlier (3.2.2) fixes $s_n = q_n$.

It remains to determine the unknown coefficients $r_n$ and $p_n$ and we may do so by applying the as yet unused Cauchy-Riemann equations along $y = 0$, (3.0.8) and (3.0.7). Using these and the standard result

$$\int_0^1 \sin m\pi x \cos n\pi x \, dx = \begin{cases} \dfrac{m\left(1 - (-1)^{n+m}\right)}{\pi \left(m^2 - n^2\right)} & n \neq m, \\ 0 & n = m, \end{cases} \tag{3.2.14}$$

it is relatively simple to show that

$$p_m = \frac{1}{\cosh m\pi} \left( q_m - \frac{2 \sinh m\pi}{\pi \mathrm{Re}} \sum_{n,\, n \neq m} n r_n \frac{\left(1 - (-1)^{n+m}\right)}{m^2 - n^2} \right), \tag{3.2.15}$$

$$r_m = \frac{1}{\cosh m\pi} \left( s_m + \frac{2\mathrm{Re} \sinh m\pi}{\pi} \sum_{n,\, n \neq m} n p_n \frac{\left(1 - (-1)^{n+m}\right)}{m^2 - n^2} \right). \tag{3.2.16}$$

These expressions for $r_m$ and $p_m$ may be investigated numerically, and we include some of our results in figure (3.11). These were calculated from a first guess of $r_n = 0$ and an iterative procedure, with both $r_n$ and $p_n$ treated as known at each stage of the working. Figures (a), (c) and (e) show results for the pressure for 50, 100 and 200 Fourier coefficients respectively, likewise in (b), (d) and (f) for vorticity. In these plots (and all similar plots for $p$, $\zeta$ which follow in this section) contours are taken at

$$[-0.1, \ -0.3, \ -0.5, \ -0.8, \ 0, \ 0.1, \ 0.2, \ 0.3, \ 0.4, \ 0.5, \ 0.6, \ 0.7, \ 0.8, \ 1], \qquad (3.2.17)$$

labelled as such in the figures, and a value of $\mathrm{Re} = 1$ is assumed throughout the calculations. We notice some difference between the plots for 50 and 100 coefficients, but little change thereafter, suggesting only 100 coefficients are needed for a complete description to the flow field. Along $y = 1$ the pressure and vorticity match the prescribed conditions, and the functions diffuse as might be expected in both cases to a value of around $p, \zeta = 0.2$ at the center of the square. Both pressure and vorticity are positive in the bottom corners. In the bottom left hand corner as $x$ increases, we see a small area of negative pressure before it once again becomes positive, peaking slightly to the right of the center of the square. In the same way as we found in the circular seed example, the figures suggest that vorticity forms a solution which is the reflection of $p$ about $x = 1/2$, and we notice an identical pattern for vorticity near $x = 1$ as $x$ decreases.

It is worth pointing out that with this coupled form of $r_m$ in (3.2.16) and $p_m$ in (3.2.15), we have been unable to find a steady $p$ and $\zeta$ with which we can express a function $F(z)$ analytically. From our work on the circular seed, this perhaps suggests that the square droplet does not yield steady velocities. Regardless we may use this series solution in validating our fully numerical approach and discuss the implications on velocity later. So, with results which seem reasonable and hold up well with increasing coefficients, we switch our focus to that task: a finite difference solution to the steady state problem.

(a) Pressure for 50 coefficients       (b) Vorticity for 50 coefficients

(c) Pressure for 100 coefficients       (d) Vorticity for 100 coefficients

(e) Pressure for 200 coefficients       (f) Vorticity for 200 coefficients

Figure 3.11: Series solution results for steady flow within the square droplet.

**A numerical algorithm**

To solve the steady state problem numerically, we can use an iterative method and labelling system on pressure and vorticity similar to the one seen in Section 2.1 for the lid driven cavity. Here $p$ and $\zeta$ are solved simultaneously via the second order accurate finite difference form of (3.0.5) and (3.0.6) (with the time derivative term equal to zero) and boundary conditions (3.2.1), (3.2.2) and (3.2.3). The procedure is terminated once the relative error between successive iterations drops below $10^{-5}$. The Cauchy-Riemann equations which act as a condition along $y = 0$, and couple the two otherwise independent systems together, are dealt with in the same way as in the boundary condition for vorticity in Section 2.1; the central difference forms of (3.0.8), (3.0.7) are used to eliminate the shadow point from (3.0.5), (3.0.6) respectively.

With this approach, it remains to solve the following system for pressure,

$$p_0 = \frac{1}{4}\left(p_1 + p_2 + p_3 + p_4\right), \tag{3.2.18}$$

$$p_0 = 1 \text{ along } x = 0, 1, \tag{3.2.19}$$

$$p_0 = \sin^2 \pi x \text{ along } y = 1, \tag{3.2.20}$$

$$p_0 = \frac{1}{4}\left(p_1 + p_3 + 2p_2 - \mathrm{Re}^{-1}\zeta_1 + \mathrm{Re}^{-1}\zeta_3\right) \text{ along } y = 0. \tag{3.2.21}$$

That is coupled with the system for vorticity,

$$\zeta_0 = \frac{1}{4}\left(\zeta_1 + \zeta_2 + \zeta_3 + \zeta_4\right), \tag{3.2.22}$$

$$\zeta_0 = 1 \text{ along } x = 0, 1, \tag{3.2.23}$$

$$\zeta_0 = \sin^2 \pi x \text{ along } y = 1, \tag{3.2.24}$$

$$\zeta_0 = \frac{1}{4}\left(\zeta_1 + \zeta_3 + 2\zeta_2 + \mathrm{Re}p_1 - \mathrm{Re}p_3\right) \text{ along } y = 0. \tag{3.2.25}$$

Some results for the converged iterations of different grid spacings are shown in figure (3.12). (a), (c) and (e) are contour plots for the pressure with 51, 101 and 201 grid points in $x$ and $y$;

(a) Pressure for 51 grid points

(b) Vorticity for 51 grid points

(c) Pressure for 101 grid points.

(d) Vorticity for 101 grid points.

(e) Pressure for 201 grid points.

(f) Vorticity for 201 grid points.

Figure 3.12: Numerically calculated results for steady flow within the square droplet.

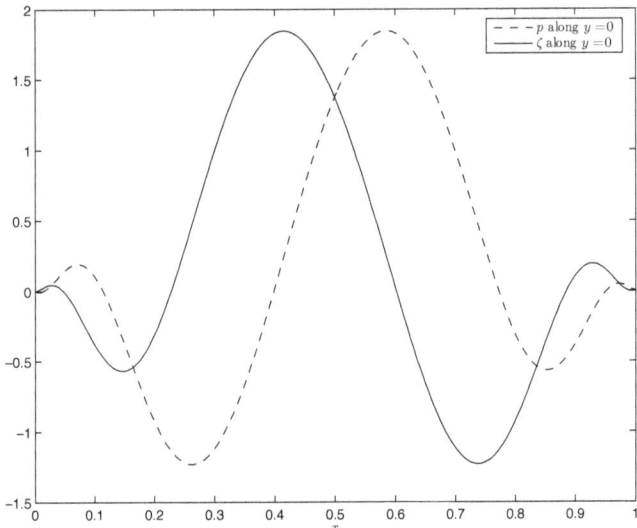

Figure 3.13: Pressure and vorticity along $y = 0$

likewise for vorticity in (b), (d) and (f).

While the plots differ slightly from those seen in figure (3.11) for the series solution, the trend seems the same in both sets, particularly when comparing the images with finer grid spacings (3.12e), (3.12f) to those with many Fourier coefficients (3.11e) and (3.11f). It was also found that our converged steady state solution was independent of the initial guess for vorticity, ($\zeta = 0$, $\zeta = \sin x$ and $\zeta$ as a computer generated random number were all tried) but we do not feel it necessary to include these plots here. We see again the pattern of oscillating pressure and vorticity close to the wall. Figure (3.13) which shows the converged $p$ and $\zeta$ along $y = 0$ for 201 grid points illustrates this, and here at least, where (3.0.8) and (3.0.7) are satisfied, $p$, $\zeta$ form a solution in which they are precisely reflections of one another about $x = 1/2$ as seen in the circular seed case. The solutions to the Cauchy-Riemann equations which form along the solid wall turn out to yield an interesting analytical problem in their own right, which we

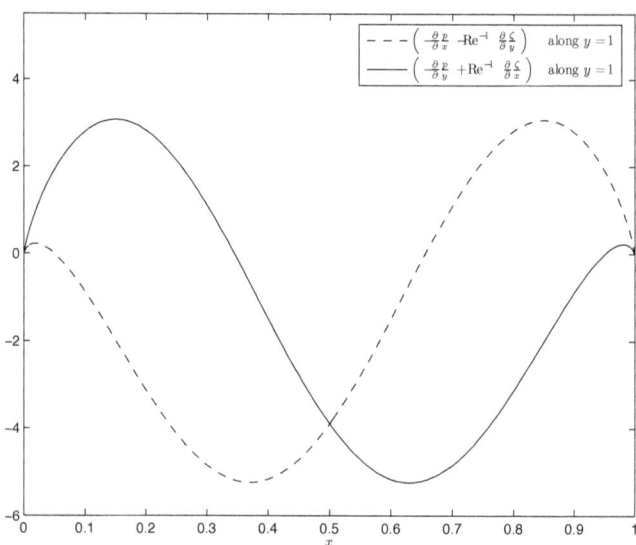

Figure 3.14: The Cauchy-Riemann-like quantities along $y = 1$ for the steady state problem of the square droplet.

will investigate further via a canonical example in Chapter 5 at the end of this part of the thesis.

We also include a plot of $(-p_x - \mathrm{Re}^{-1}\zeta_y)$ and $(-p_y + \mathrm{Re}^{-1}\zeta_x)$ along $y = 1$ in figure (3.14) (where subscript denotes differentiation), clearly demonstrating that, unlike the circular seed example, for the square droplet the Cauchy-Riemann-like terms do not sum to zero. The trend is similar throughout the droplet for $0 < y \leq 1$.

Taking (3.0.1) and (3.0.2) into account, this has the direct result of making the velocities grow linearly with time. If our time marching procedure does converge to the unique steady $p$ and $\zeta$ shown here, there will not be a unique velocity field attached to the solution and we expect to find velocities which follow the trend

$$u \sim t \left( -\frac{\partial p}{\partial x} - \mathrm{Re}^{-1} \frac{\partial \zeta}{\partial y} \right), \tag{3.2.26}$$

$$v \sim t \left( -\frac{\partial p}{\partial y} + \mathrm{Re}^{-1} \frac{\partial \zeta}{\partial x} \right), \tag{3.2.27}$$

where the terms in the brackets of course are constant in time and non zero. The velocities in

this form would suggest

$$\frac{\partial u}{\partial x} + \frac{\partial v}{\partial y} = 0, \tag{3.2.28}$$

$$\frac{\partial u}{\partial y} - \frac{\partial v}{\partial x} = 0; \tag{3.2.29}$$

indeed, the first of these is the continuity equation, and the second may also be derived from

the leading order terms in the steady vorticity. These are the Cauchy-Riemann equations for

velocity, directly implying that both $u$ and $v$ are harmonic functions which satisfy Laplace's

equation and that we may define an analytic function,

$$F(z) = v + iu. \tag{3.2.30}$$

This analytic function $F(z)$ cannot be zero along a segment of any curve however, suggesting

that to satisfy the no slip condition, $u$ and $v$ are inviscid to leading order but form a boundary

layer near $y = 0$.

Within the core, $F(z)$ will be non-unique, but will be a linear scale factor of the $u, v$ solu-

tion at any given time step once pressure and vorticity have become steady; thus a given $u$

yields a unique solution for $v$ and $F$. Applying the small ratios parameter to the shape of the

droplet (initially $y = 1$ in this canonical problem) so that it is described by

$$f = f_0 + \epsilon f_1 + \cdots, \tag{3.2.31}$$

in the early stage, the kinematic condition suggests the $O(1)$ contribution to shape $f_0$ is time

independent and the perturbation may be found from

$$\frac{\partial f_1}{\partial t} = v - u \frac{\partial f_0}{\partial x}. \tag{3.2.32}$$

In our case of course, $f_0$ is unity and the last term in (3.2.32) is zero. If $v$ grows linearly with time we expect to see $f_1$ growing like time squared. Thus far, our results seem to tie in well analytically and numerically but are surprising perhaps in light of the circular seed results. To validate the above hypothesis about the velocities, we must confirm that the full unsteady problem does converge to this solution at large times.

### 3.2.2 The time dependent problem

The algorithm we use for our time marching procedure is similar to those seen earlier in the thesis (lid driven cavity, circular seed and so on). We apply a first order accurate backwards finite difference scheme in time and a second order accurate central difference scheme in space to the full form of (3.0.5), (3.0.6) and boundary conditions (3.2.1) (3.2.2). The final condition along $y = 0$ is the same as that used in the steady state problem. With this approach, we aim to numerically solve the following system for pressure

$$p_0 = \frac{1}{4}\left(p_1 + p_2 + p_3 + p_4\right), \tag{3.2.33}$$

$$p_0 = 1 \text{ along } x = 0, 1, \tag{3.2.34}$$

$$p_0 = g(t)\sin^2 \pi x \text{ along } y = 1, \tag{3.2.35}$$

$$p_0 = \frac{1}{4}\left(p_1 + p_3 + 2p_3 - \text{Re}^{-1}\zeta_1 + \text{Re}^{-1}\zeta_3\right). \tag{3.2.36}$$

Coupled with the system for vorticity

$$\zeta_0 = \frac{\delta t}{\text{Re}h^2 + 4\delta t}\left(\zeta_1 + \zeta_2 + \zeta_3 + \zeta_4\right) + \frac{\text{Re}h^2}{\text{Re}h^2 + 4\delta t}\zeta_0^{OLD}, \tag{3.2.37}$$

$$\zeta_0 = 1 \text{ along } x = 0, 1, \tag{3.2.38}$$

$$\zeta_0 = g(t)\sin^2 \pi x \text{ along } y = 1, \tag{3.2.39}$$

$$\zeta_0 = \frac{\delta t}{\text{Re}h^2 + 4\delta t}\left(\zeta_1 + \zeta_3 + 2\zeta_2 + \text{Re}p_1 - \text{Re}p_3\right) + \frac{\text{Re}h^2}{\text{Re}h^2 + 4\delta t}\zeta_0^{OLD}, \tag{3.2.40}$$

where the subscript refers to the 5 point stencil labelling seen earlier and $OLD$ is the value of $\zeta$ at the previous time step. Starting with an initial condition of $\zeta = p = 0$ everywhere, we use an iterative procedure identical to that in the steady case to solve for the first time step. Once this solution has converged, we update $\zeta^{OLD}$, find the velocities from (3.0.1), (3.0.2) and the

shape effect $f_1$ from the kinematic condition (3.2.32). The iterative process is then repeated for

subsequent time steps.

It was found that no relaxation or negative initial time (see section (3.1)) were required. Some

results at times $t = 0.5, 1, 2, 4, 8, 16$, for a grid with $h = 1/200$, 201 points in both $x$ and $y$

and a time step of $\delta t = h^2/2$, are shown in figure (3.15) for pressure and (3.16) for vorticity.

We see from the plots that the pattern of oscillating pressure and vorticity values along $y = 0$

appears early in the simulation, increasing in magnitude and diffusing through the droplet as

time increases. The system appears to become virtually steady by around $t = 8$ with little

change between (e) and (f) in both figures. Crucially, we also notice the similarities between

the steady state images found from the time marching algorithm, (3.15f) and (3.16 f), and those

from the steady state problem (3.12e) and (3.12f). All contour values and general trends match

well.

We also include plots of the time marched velocities along $y = 1$ in figure (3.17), found

from the integration of (3.0.1) and (3.0.2) at each time step. We notice, as suggested in the

previous section, and contrary to the circular seed case, that the velocities do indeed grow

linearly with time, at least at relatively large times. For instance $u(t = 16)$ has a maximum

value of around 45, while $\max(u(t = 8)) \approx 23$. The story is the same in the image for $v$.

Importantly, the shape of solutions matches exactly those seen in figure (3.14): $u$ and $v$ at any

large $t$ do indeed turn out to be a constant multiple of $-(p_x + \text{Re}^{-1}\zeta_y)$ and $(-p_y + \text{Re}^{-1}\zeta_x)$

respectively.

Finally, we examine the shape effect $f_1$ described in (3.2.31). Plots at different times are

shown in figure (3.18). At each time step $f_1$ is found from the kinematic condition (3.2.32).

The fluid within the droplet fills the area of low pressure at the left hand side of the interface and

(a) $t = 1/2$

(b) $t = 1$

(c) $t = 2$

(d) $t = 4$

(e) $t = 8$

(f) $t = 16$

Figure 3.15: Time marched results for pressure $p$

(a) $t = 1/2$          (b) $t = 1$

(c) $t = 2$          (d) $t = 4$

(e) $t = 8$          (f) $t = 16$

Figure 3.16: Time marched results for vorticity $\zeta$

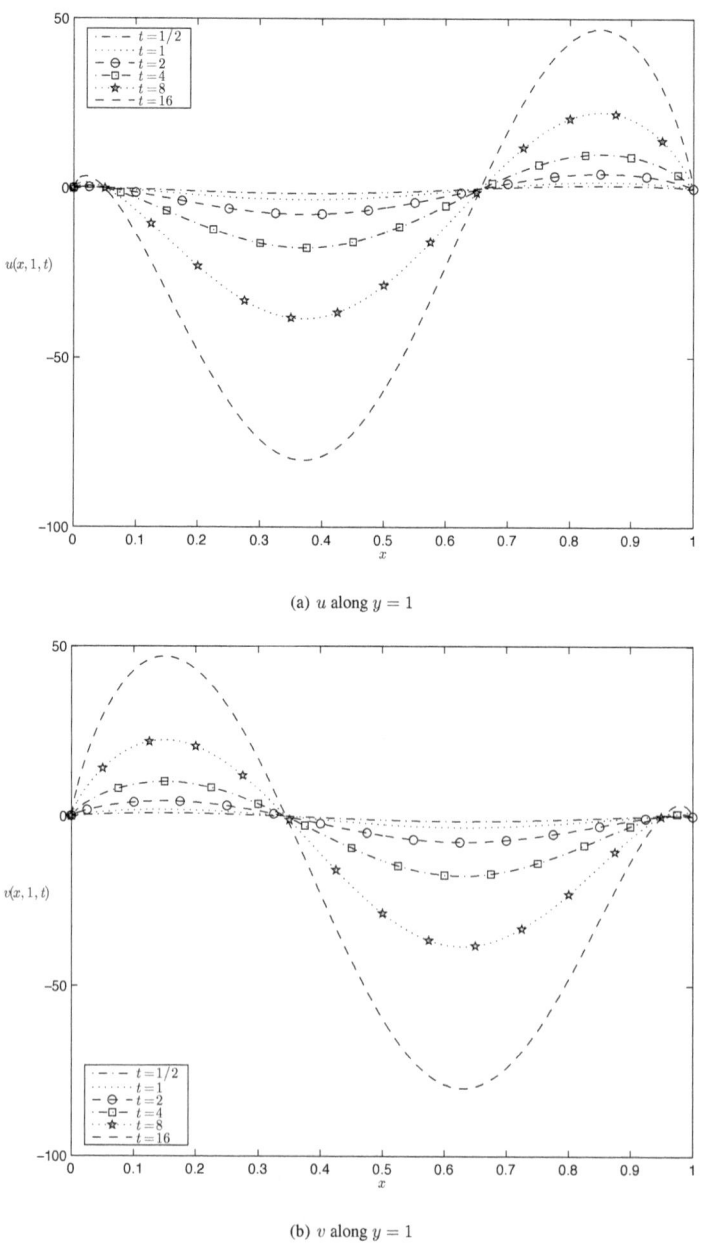

(a) $u$ along $y = 1$

(b) $v$ along $y = 1$

Figure 3.17: The time marched velocities along the top of the droplet

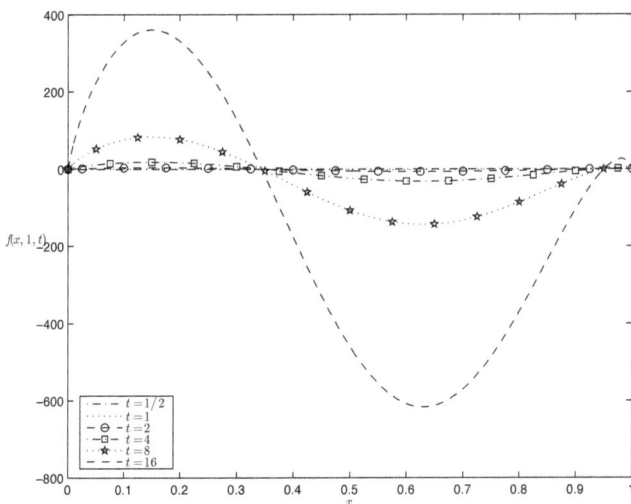

Figure 3.18: The shape effect $f_1$ of the square droplet plotted against $x$ for different times

is then forced downwards as pressure increases to its maximum. Considering the movement seen in real droplets, this perhaps may not seem reasonable, but we bear in mind that neither our interface condition nor the droplet geometry are designed to be realistic yet. The main conclusion may still be drawn from this image: that the shape effect grows like time squared.

It is worth noting that a series solution for the time dependent problem could be achieved to validate these results further, but our earlier choice of prescribed condition, (3.2.1) and (3.2.2), makes this a non-trivial task. A different choice (say the sum of a steady and exponentially decaying time dependent term) would simplify the process, but since our motivations lie only in the insight the square droplet offers us for the semicircular one, and our time marching algorithm is already validated by its similarities to the steady state problem, we choose not to explore this avenue.

With two examples contradicting in terms of their large time behaviour, we move on to the main focus of this chapter: flow within a semicircular droplet.

## 3.3 The semicircular droplet

Our work in the previous sections of this chapter has been motivated by this semicircular droplet problem, corresponding to step two of our method for modelling droplet deformation. An overview of the method may be found in Chapter 1 along with a diagram of the flow regime, figure (1.1), and the system of equations which govern the flow (1.2.15), (1.2.16).

The problem of Stokes or creeping flow within a semicircle has been tackled analytically by other authors. Melesko and Gomilko [34], for example, present an exact solution for motion induced by a constant tangential velocity along the boundary. Our wider aim however is to connect a solution within the droplet to the numerical, time dependent results for pressure and vorticity which were found in Chapter 2. As such, rather than approach the problem analytically, we require a general numerical algorithm to solve for flow within the semicircle.

To do so, we begin as we did in the earlier sections of this chapter by prescribing idealised $p$ and $\zeta$ along the free surface of the droplet:

$$p(1, \theta) = -\frac{e^t}{1 + e^t} \sin^2 \theta, \tag{3.3.1}$$

$$\zeta(1, \theta) = \frac{e^t}{1 + e^t} \sin \theta. \tag{3.3.2}$$

so that a complete description to the problem is given in pressure-vorticity form by (3.0.5) and (3.0.6) subject to (3.3.1), (3.3.2) along the interface and a no slip condition - equivalent to the Cauchy-Riemann equations - along $y = 0$, (3.0.14) and (3.0.15).

As in previous sections, the inclusion of a time dependent term in (3.3.1), (3.3.2) allows for a gentle introduction of the interface conditions, avoiding Rayleigh layers and the numerical difficulties associated with them.

The Stokes equations in this form might cause a singularity in both pressure and vorticity

at $r = 0$, despite the fact the points $(\delta r, 0)$ and $(\delta r, \pi)$ are adjacent to one another in a Cartesian

sense. This issue was avoided in the circular seed example due to the solid center for $0 \leq r \leq a$.

To avoid this problem here, we rewrite the governing equations in terms of two new variables:

$$q = rp, \quad z = r\zeta, \tag{3.3.3}$$

and fix $q = z = 0$ along $r = 0$. With this approach, it remains to solve the following system

for $q$:

$$\frac{\partial^2 q}{\partial \theta^2} - r\frac{\partial q}{\partial r} + r^2\frac{\partial^2 q}{\partial r^2} + q = 0, \tag{3.3.4}$$

$$q = 0 \text{ along } r = 0, \tag{3.3.5}$$

$$q = -\frac{e^t}{1 + e^t}\sin^2\theta \text{ along } r = 1, \tag{3.3.6}$$

$$\frac{\partial q}{\partial \theta} = \text{Re}^{-1}\left(r\frac{\partial z}{\partial r} - z\right) \text{ along } \theta = 0, \pi \tag{3.3.7}$$

coupled with the system for $z$:

$$r^2\frac{\partial z}{\partial t} = \text{Re}^{-1}\left(r^2\frac{\partial^2 z}{\partial r^2} - r\frac{\partial z}{\partial r} + z + \frac{\partial^2 z}{\partial \theta^2}\right), \tag{3.3.8}$$

$$z = 0 \text{ along } r = 0, \tag{3.3.9}$$

$$z = -\frac{e^t}{1 + e^t}\sin\theta \text{ along } r = 1, \tag{3.3.10}$$

$$\frac{\partial z}{\partial \theta} = \text{Re}\left(q - r\frac{\partial q}{\partial r}\right) \text{ along } \theta = 0, \pi. \tag{3.3.11}$$

The full form of $p$ and $\zeta$ may then be found once the numerical procedure is complete, us-

ing (3.3.3) and applying a linear interpolation between the two adjacent points $p(\delta r, \pi)$ and

$p(\delta r, 0)$, to reconstruct the true value of pressure at $r = 0$. An identical method finds $\zeta(0, \theta)$.

The algorithm appears to handle this method well as may be seen in our later plots.

An additional difficulty arose in achieving finer grid spacings in the radial direction. We suspect this may be due to the singularity avoidance at $r = 0$ and our use of (3.3.3). However, given the time restraints of the project, and that the values of $\delta r$ for which the time dependent solution did converge appear to capture the dynamics of the droplet well, we did not think it necessary to adapt our numerical procedure to consider this matter further. Our simulations did not highlight any such problem in the azimuthal direction, so, in order to retain resolution in $\theta$ we define our grid spacings with $\delta r \neq \delta \theta$ and set $N_r$ and $N_\theta$ as the number of grid points in $r$ and $\theta$ respectively.

Despite their different large time behavior, the flow simulations in both the circular seed and the square droplet achieved a steady state pressure and vorticity by approximately $t = 8$. We suggest, or anticipate that the semicircular droplet will also and so, since the time marching algorithm turns out to be computationally expensive, choose to terminate out calculations at this time point. To help further, we choose to apply the Taylor's expansion of the two points immediately interior to the boundary (as in (2.2.15)) rather than the shadow point elimination seen in this context previously. We include in figure (3.19) an example of our labelling system for this method at the boundary $\theta = 0$. With the points referenced in this way, the finite difference form of $\dfrac{\partial q}{\partial \theta}$ along $\theta = 0$ is written:

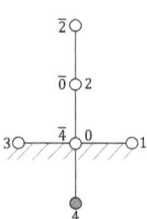

Figure 3.19: Labelling system along $\theta = 0$.

$$\left.\frac{\partial q}{\partial \theta}\right|_{\theta=0} = -\frac{(q_{\bar{2}} - 4q_{\bar{0}} + 3q_{\bar{4}})}{2\delta\theta}, \tag{3.3.12}$$

and the same approach is used along $\theta = \pi$.

Aside from these features, the labelling system and difference quotients for (3.3.4) to (3.3.11) are the same as those outlined in the lid driven cavity example in section (2.1), leading us to the system which governs our numerical algorithm. Equations (3.3.7) and (3.3.4) are replaced by:

$$q_{\bar{4}} = \frac{4q_{\bar{0}}}{3} - \frac{q_{\bar{2}}}{3} + \frac{2\delta\theta}{3\mathrm{Re}}z_0 + \frac{r_0\delta\theta}{3\mathrm{Re}\delta r}z_3 - \frac{r_0\delta\theta}{3\mathrm{Re}\delta r}z_1 \quad \text{along } \theta = 0, \tag{3.3.13}$$

$$q_{\bar{2}} = \frac{4q_{\bar{0}}}{3} - \frac{q_{\bar{4}}}{3} - \frac{2\delta\theta}{3\mathrm{Re}}z_0 - \frac{r_0\delta\theta}{3\mathrm{Re}\delta r}z_3 + \frac{r_0\delta\theta}{3\mathrm{Re}\delta r}z_1 \quad \text{along } \theta = \pi, \tag{3.3.14}$$

$$q_0 = \left\{ \frac{r_0}{2}\left(2r_0 - \delta r\right)q_1 + \left(2r_0 + \delta r\right)q_3 + \frac{\delta r^2}{\delta\theta^2}\left(q_2 + q_4\right) \right\} \Big/ \left( 2r_0^2 + \frac{2\delta r^2}{\delta\theta^2} - \delta r^2 \right), \tag{3.3.15}$$

and (3.3.11), (3.3.8) by:

$$z_{\bar{4}} = \frac{4z_{\bar{0}}}{3} - \frac{z_{\bar{2}}}{3} - \frac{2\delta\theta}{3\mathrm{Re}}q_0 - \frac{r_0\delta\theta}{3\mathrm{Re}\delta r}q_3 + \frac{r_0\delta\theta}{3\mathrm{Re}\delta r}q_1 \quad \text{along } \theta = 0, \tag{3.3.16}$$

$$z_{\bar{2}} = \frac{4z_{\bar{0}}}{3} - \frac{z_{\bar{4}}}{3} + \frac{2\delta\theta}{3\mathrm{Re}}q_0 + \frac{r_0\delta\theta}{3\mathrm{Re}\delta r}q_3 - \frac{r_0\delta\theta}{3\mathrm{Re}\delta r}q_1 \quad \text{along } \theta = \pi, \tag{3.3.17}$$

$$z_0 = \left\{ \frac{r_0}{2}\left(2r_0 - \delta r\right)z_1 + \left(2r_0 + \delta r\right)z_3 + \frac{\delta r^2}{\delta\theta^2}\left(z_2 + z_4\right) \right.$$
$$\left. + \frac{r_0^2\mathrm{Re}\delta r^2}{\delta t}z_0^{old} \right\} \Big/ \left( 2r_0^2 + \frac{2\delta r^2}{\delta\theta^2} + \frac{r_0^2\mathrm{Re}\delta r^2}{\delta t} - \delta r^2 \right). \tag{3.3.18}$$

The method used to solve equations (3.3.13) to (3.3.18) (alongside equivalent forms for (3.3.5), (3.3.6), (3.3.9), (3.3.10) of course) is identical to those used repeatedly throughout the chapter, details of which we do not think necessary to include here.

As mentioned, we expect the unsteady problem to converge in time to a steady state. This steady state may be modelled separately by considering (3.3.4) to (3.3.11) with all time derivative terms set to zero and $\frac{e^t}{1+e^t}$ taken as unity. We present the results for the pressure and vorticity along $y = 0$ found via the two separate methods together in figure (3.20). The left hand column in the figure shows solutions obtained from the steady state algorithm for grids with $(N_r, N_\theta) = (41, 151), (51, 201), (61, 251)$. Those in the right hand column are snapshots of the time marched $p$, $\zeta$ taken at $t = 8$.

The plots demonstrate good agreement between the two sets suggesting we do indeed see $p$ and $\zeta$ becoming steady at large times. Here it becomes clear that a linear interpolation between the two Cartesian-adjacent points determines the finite values of $q/r$ and $z/r$ at $r = 0$

well, with no discontinuity or jump near this point. The trend of $p$ and $\zeta$ along the wall, where

the Cauchy-Riemann equations apply, matches well to the analogous results for the square

droplet seen in figure (3.13). In the current model, $p$ is no longer an exact reflection of $\zeta$ about

$x = 0$ or $\theta = \pi/2$, but this is due to the difference between (3.3.1) and (3.3.2): the chosen

interface conditions for $p$ and $\zeta$ were identical in the square droplet example.

We notice the magnitude of $p$ and $\zeta$ along $y = 0$ seems somewhat sensitive to grid spac-

ing for these values of $\delta r$ and to investigate this further we include contour plots of both at

different grid spacings and a range of times. Those for pressure may be seen in figure (3.21) and

for vorticity in (3.22). In the left hand column of both figures, the solution is found from a grid

of $(N_r, N_\theta) = (51, 201)$ and plots are shown at $t = 0.5, 1, 2, 4, 8$. To allow a comparison, the

right hand column shows the same but with $(N_r, N_\theta) = (61, 251)$. In both figures, contours for

pressure and vorticity respectively are plotted at the same values given in (3.2.17), and labelled

as such in the figures.

At smaller times and close to the origin $r = 0$, we notice some difference between the so-

lutions calculated on the two grids. However, as $r \to 1$ in all images, and throughout the flow

geometry as $t$ becomes larger, the two sets generally agree. The overall trend is the same and

the contour labels match well.

While the restrictions in $\delta r$ limit the accuracy of our results, we conclude that the treatment is

still able to capture the heart of the solution and we move on to investigate the behavior of the

velocities once $p$ and $\zeta$ have reached a steady state.

If the velocities also become steady, as they did in the example of the circular seed, the steady

state $p$, $\zeta$ will form solutions to the Cauchy-Riemann equations everywhere and a function

$F(z) = \zeta + ip$ will be analytic in $z$.

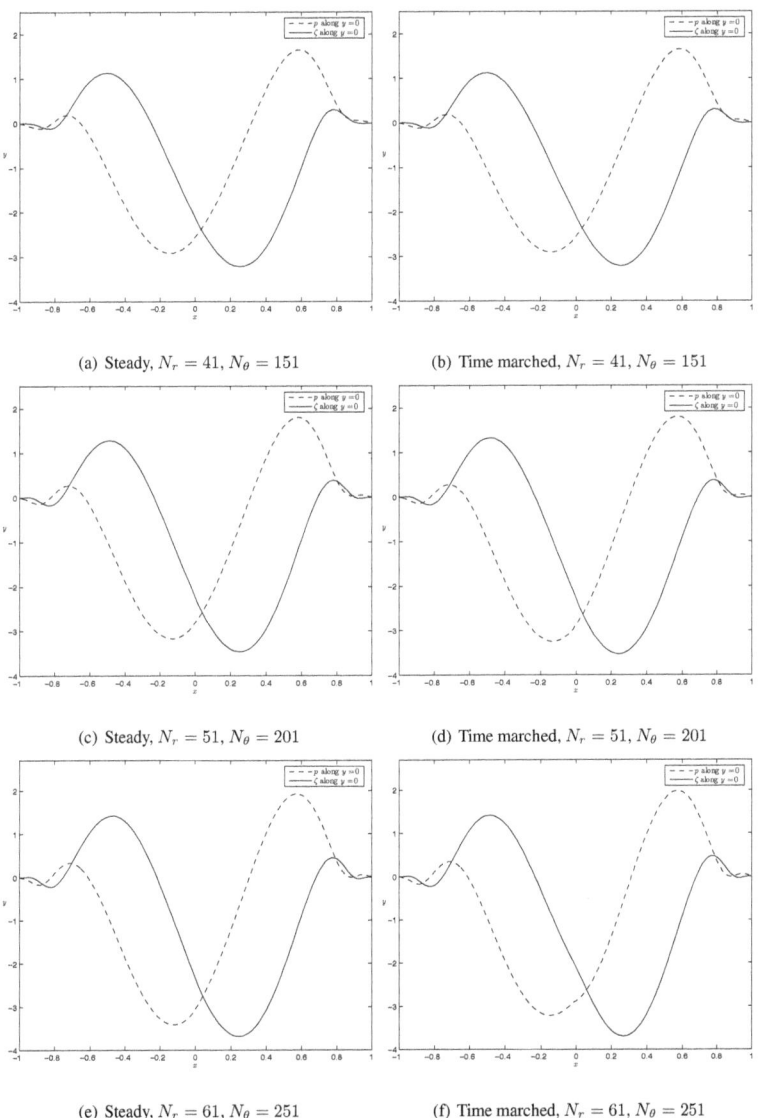

(a) Steady, $N_r = 41$, $N_\theta = 151$

(b) Time marched, $N_r = 41$, $N_\theta = 151$

(c) Steady, $N_r = 51$, $N_\theta = 201$

(d) Time marched, $N_r = 51$, $N_\theta = 201$

(e) Steady, $N_r = 61$, $N_\theta = 251$

(f) Time marched, $N_r = 61$, $N_\theta = 251$

Figure 3.20: The pressure and vorticity against $x$ along $y = 0$ found via the time marching algorithm (plotted at $t = 8$) and the steady state problem. Various grid spacings are included.

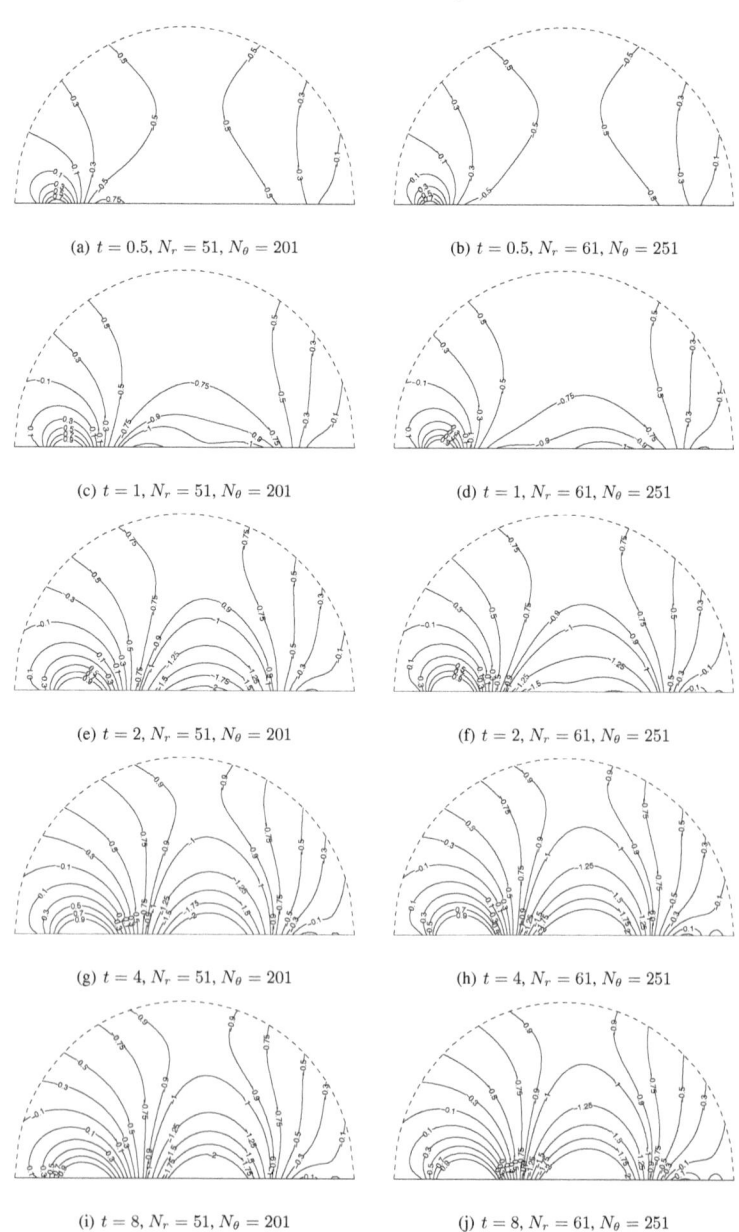

(a) $t = 0.5$, $N_r = 51$, $N_\theta = 201$        (b) $t = 0.5$, $N_r = 61$, $N_\theta = 251$

(c) $t = 1$, $N_r = 51$, $N_\theta = 201$        (d) $t = 1$, $N_r = 61$, $N_\theta = 251$

(e) $t = 2$, $N_r = 51$, $N_\theta = 201$        (f) $t = 2$, $N_r = 61$, $N_\theta = 251$

(g) $t = 4$, $N_r = 51$, $N_\theta = 201$        (h) $t = 4$, $N_r = 61$, $N_\theta = 251$

(i) $t = 8$, $N_r = 51$, $N_\theta = 201$        (j) $t = 8$, $N_r = 61$, $N_\theta = 251$

Figure 3.21: Grid refinement example for lines of constant pressure within the semicircular droplet.

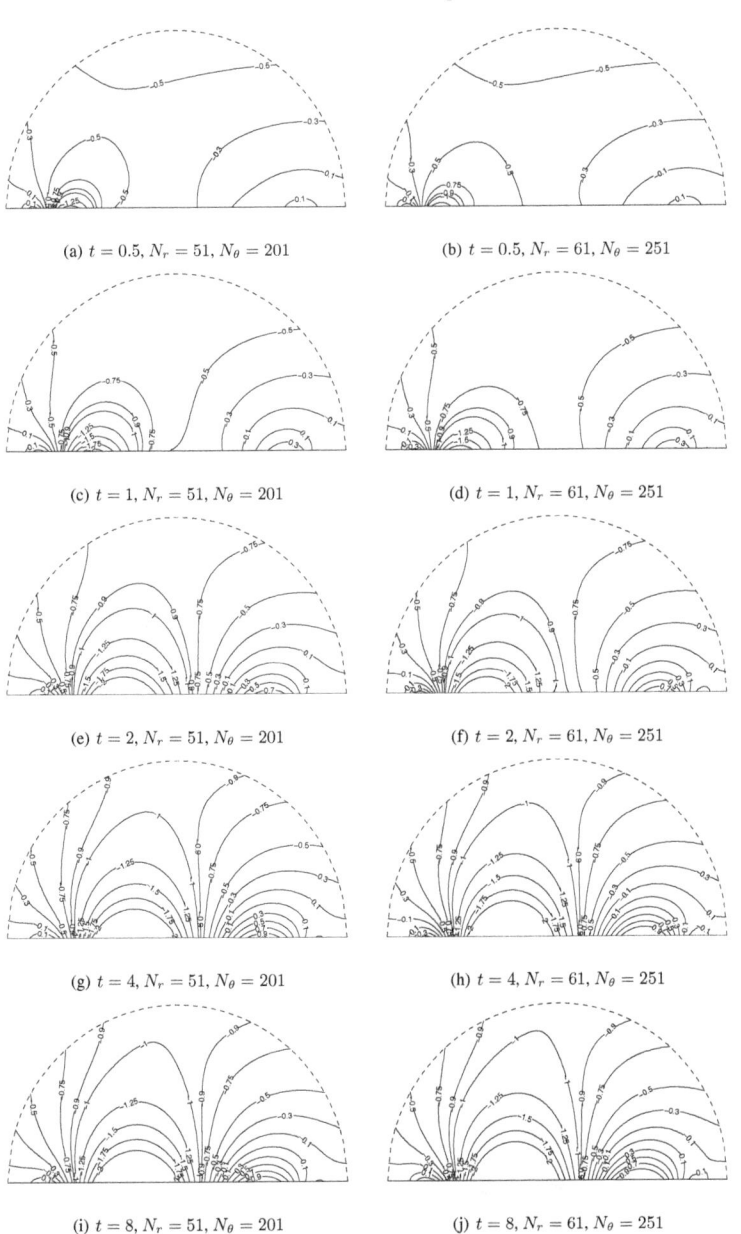

(a) $t = 0.5$, $N_r = 51$, $N_\theta = 201$

(b) $t = 0.5$, $N_r = 61$, $N_\theta = 251$

(c) $t = 1$, $N_r = 51$, $N_\theta = 201$

(d) $t = 1$, $N_r = 61$, $N_\theta = 251$

(e) $t = 2$, $N_r = 51$, $N_\theta = 201$

(f) $t = 2$, $N_r = 61$, $N_\theta = 251$

(g) $t = 4$, $N_r = 51$, $N_\theta = 201$

(h) $t = 4$, $N_r = 61$, $N_\theta = 251$

(i) $t = 8$, $N_r = 51$, $N_\theta = 201$

(j) $t = 8$, $N_r = 61$, $N_\theta = 251$

Figure 3.22: Grid refinement example for lines of constant vorticity within the semicircular droplet.

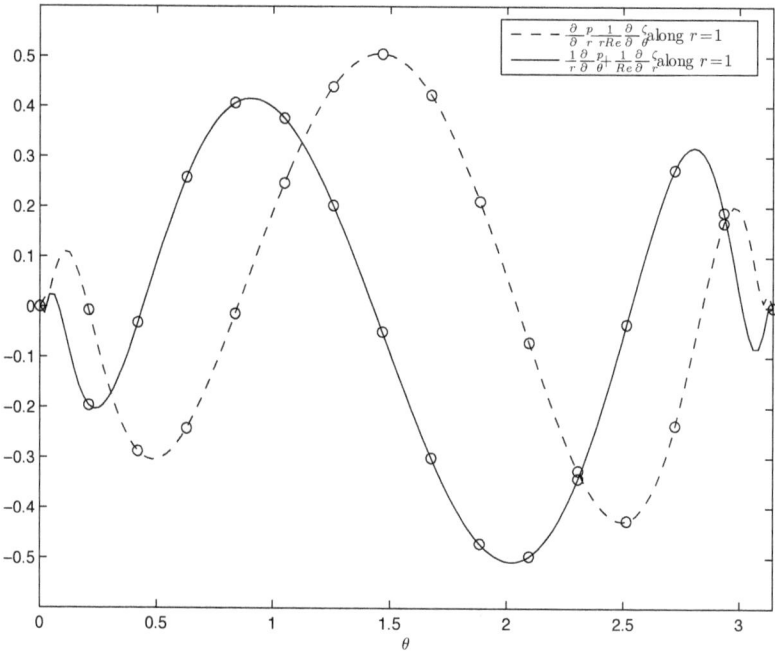

Figure 3.23: The Cauchy-Riemann-like quantities against $\theta$ and along $r = 1$ found from the time marching algorithm demonstrate that (3.0.14) and (3.0.15) are not satisfied. The results from the steady state problem are also included and plotted as circles.

On the other hand, a finding of velocities growing linearly with time, as seen in the square droplet, implies that $p$, $\zeta$ will satisfy the Cauchy-Riemann equations only along $y = 0$. Instead, $u_r$, $u_\theta$ become harmonic.

Figure (3.23), a plot of the Cauchy-Riemann type terms along $r = 1$ gives us the answer. These results are calculated on a grid with $(N_r, N_\theta) = (61, 251)$, and the curve is drawn at $t = 8$, clearly demonstrating that (3.0.14) and (3.0.15) are not satisfied here: the velocities within the semicircular droplet form solutions of the type seen in the square droplet and will

grow linearly with time. Included in the figure, and shown as circles, are the same terms found

from the steady state algorithm, demonstrating that by $t = 8$ the solution has indeed reached a

steady state and suggesting the steady state solution is independent of path.

The results displayed in figure (3.23) also form the basis of the velocities $u_r$ and $u_\theta$:

$$u_r \sim t \left( -\frac{\partial p}{\partial r} - \frac{\mathrm{Re}^{-1}}{r} \frac{\partial \zeta}{\partial \theta} \right), \tag{3.3.19}$$

$$u_\theta \sim t \left( -\frac{1}{r} \frac{\partial p}{\partial \theta} + \mathrm{Re}^{-1} \frac{\partial \zeta}{\partial r} \right), \tag{3.3.20}$$

and can be seen to be true in figures (3.24a) and (3.24b), showing $u_r$ and $u_\theta$ against $\theta$ respec-

tively along $r = 1$ for $t = 0.5, 1, 2, 4, 8$, where curves match the trend of those seen in (3.23).

We notice here that the velocities do indeed present linear growth in time. For example, the

radial velocity has a maximum interface value of around $u_r = 1.9$ at $t = 4$ compared to a

maximum of $u_r = 3.9$ at $t = 8$.

Returning to the $O(\epsilon)$ balance of the kinematic condition in polar form, (3.1.25), where the

droplet shape is taken as

$$f(\theta, t) = f_0(\theta) + \epsilon f_1(\theta, t) + \cdots, \tag{3.3.21}$$

and $f_0 = 1$ in our example, we expect $f_1$ to evolve as

$$\frac{\partial f_1}{\partial t} = u_r, \tag{3.3.22}$$

and hence $f_1$ will grow like time squared in this model. We include a plot of $f_1$ in figure (3.25a)

which demonstrates that this is indeed the case.

For clarity and completeness, we also include a Cartesian plot of the droplet shape change

in figure (3.25b). Here, the solid black line is the original shape and the free surface of the

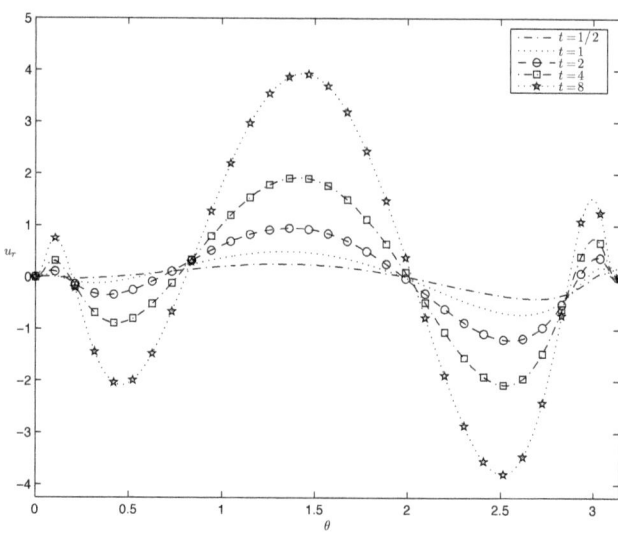

(a) Radial velocity $u_r$ against $\theta$ along the two fluid interface $r = 1$

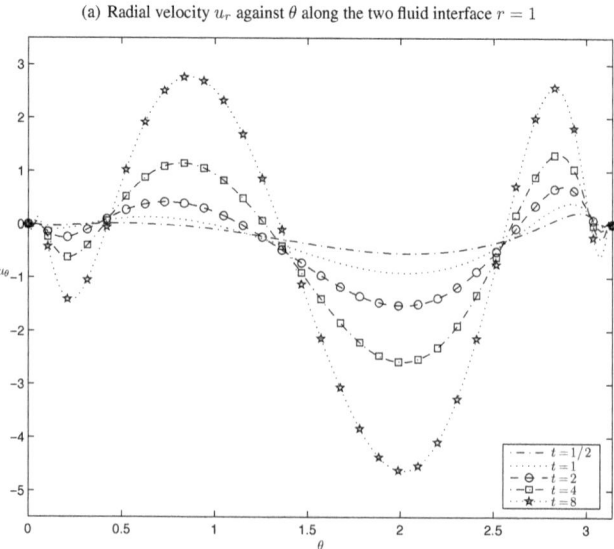

(b) Azimuthal velocity $u_\theta$ against $\theta$ along the two fluid interface $r = 1$

Figure 3.24: The velocities along the interface as time increases, demonstrating linear growth with time.

droplet is drawn at $t = 8$ for $\epsilon = 1/100, 1/200, 1/400$. Considering our initial interface conditions (3.3.1), (3.3.2), these results appear to make intuitive sense. The fluid within the droplet moves into the area of low pressure near the central station $x = 0$ and is offset to the right by the shear flow in air. Similar droplet shapes may be seen in the next part of the thesis for static droplets contained within a triple deck structure.

To achieve more realistic results, we must abandon our idealised interface conditions and return to those found in the previous chapter: combining the two algorithms to achieve an interacting system. This is our task for the next chapter, where we also include a discussion of the limitations of our model, an explanation of the anomalous results for the circular seed case and some analysis of the later temporal stage, when the two fluid interaction becomes fully nonlinear.

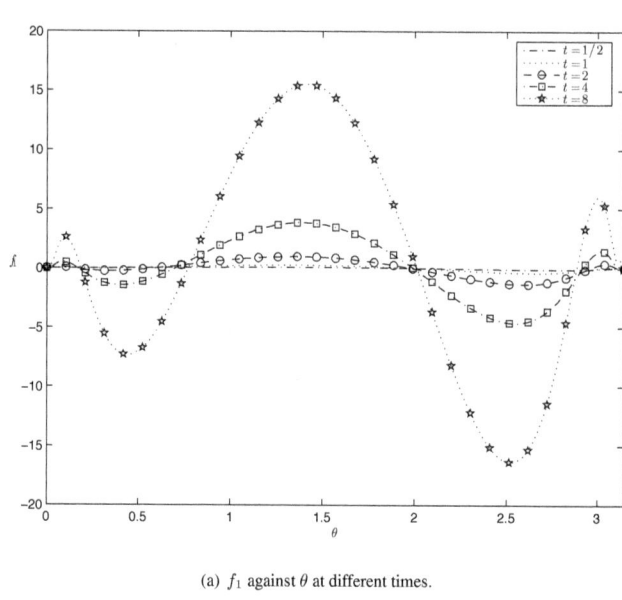

(a) $f_1$ against $\theta$ at different times.

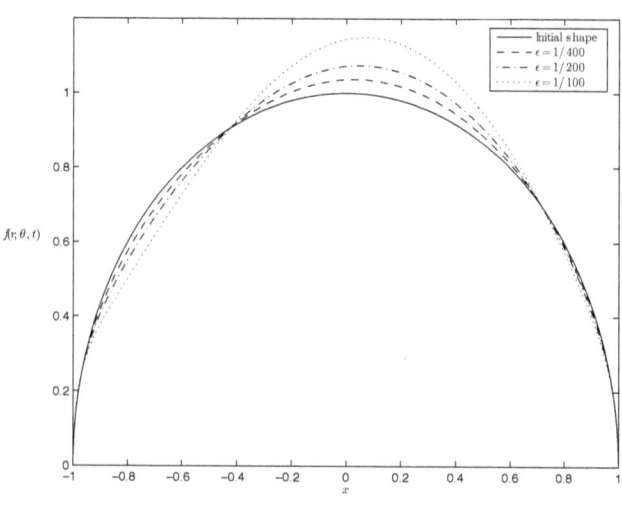

(b) Cartesian plot of shape change for chosen values of $\dfrac{1}{\epsilon}$

Figure 3.25: The evolution of the interface for the semicircular droplet.

## Chapter 4

# The interacting system

By isolating the flows of air and water, we have been able to create two numerical algorithms which separately describe the flow over a semicircular object (Chapter 2) and the flow within a semicircular droplet (Chapter 3). The focus of this chapter then, is to combine these two algorithms to achieve an interacting description of droplet deformation as outlined in the overview of the method in Chapter 1. A description of the flow regime and a diagram of the geometry is also given there.

Combining the two algorithms is relatively straightforward. Since the model describes the early temporal stage - while $\epsilon f_1 << O(1)$ and the droplet shape is semicircular to leading order - the flow in the air remains as before in Chapter 2 and the system interacts in one direction only. Taking the surface of the droplet as quasi-steady, the solution in the air is used to drive the flow in the water at each time step, but the procedure to determine the flow within the water droplet of Chapter 3 is otherwise unchanged.

To allow for the interaction, a matrix of discrete time dependent pressure and vorticity values along $r = 1$ is extracted from the numerical procedure in the air. Since a no-slip condition applies in the air, the pressure may be found from the second order accurate discrete form of

$$p = \mathrm{Re}^{-1} \int_0^\theta \left. \frac{\partial \zeta(r,s)}{\partial r} \right|_{r=1} ds, \tag{4.0.1}$$

by manipulating the Taylor expansions of $\zeta$ about $r = 1$ of the two points immediately interior

to the boundary (in air) and applying the trapezium rule to the result, using $p(1,0) = 0$ as a

reference point. This matrix then replaces the idealised interface conditions (3.3.1) and (3.3.2)

in the model for the flow within the water droplet.

This approach dictates (or at least suggests initially) that the discretised grids in air and water

must match along the boundary between the two fluids and, in our case, raises a potential

restriction on the resolution of any solution. In the flow in air model of Chapter 2 we demon-

strated that the wall at $r = r_\infty$, finite in a numerical sense, has to be far enough from the

semicircle so as to not cause an upstream effect. This and our choice to apply a uniform grid

discretisation made fine resolutions computationally expensive. Since our grid in water must

match that in air, we find the same problem here also.

It would be possible of course to return to our flow in air model, define $\delta r \neq \delta \theta$ and apply a

multi grid method to increase the number of points along the air-water boundary. However, it

appears that instead extending the model of Chapter 2 with 151 grid points along the boundary

to one of 201 points using linear interpolation works well and provides apparently reliable

results.

Aside from the restrictions surrounding resolution and computational expense, this chapter

also discusses some of the fundamental limitations in the model: the break down at the later

temporal stage (Section 4.2) and the effect of not including the full stress conditions along the

interface (Section 4.3).

While these limitations restrict the application of this particular model to a real world droplet,

we have found a special case: a droplet contained within a triple deck structure, for which we

have successfully modelled the two way nonlinear interaction between air and water associated with the second temporal stage, with full stress conditions to leading order along the interface. This is the main motivation of the second part of the thesis and numerical results and simulations are given there.

As the triple deck droplet captures so well the two-way dynamics of a real world droplet deformation we choose to focus our time and attention there, rather than in refining the interacting version of the current model. Despite this, the results which follow match closely to those found earlier in the present part of the thesis and provide a good sense of the underlying dynamics. No further adaptations to the procedures already described were required and so we progress directly to our results.

## 4.1 Results

To allow for comparison, we present the same set of figures for the current solution as we did in the case of the idealised interface problem in Chapter 3. Of course, since this interaction is based on an unchanged simulation of flow in air, we do not include those figures here, although they may be seen in Chapter 2. All calculations are based on a grid $(N_r, N_\theta) = (51, 201)$, $\mathrm{Re} = \widehat{\mathrm{Re}} = 3$ and an initial time of $t = -5$ to avoid Rayleigh layers in both fluids.

The pressure and vorticity contours are shown in figure (4.1). The left and right hand columns show the results for $p$ and $\zeta$ respectively. The system, initially at rest, yields fairly large values as time increases and for that reason, a fixed set of contours did not seem appropriate. Instead a range of values is plotted: between $(-20, 20)$ for $p$ and $(-100, 30)$ for $\zeta$. All contours are labelled accordingly in the figures.

We notice that the pressure appears largely independent of $y$ in the early stages, aside from a small area of high pressure near $r = 1$, $\theta = \pi$ and an area of low pressure near $r = 1$, $\theta = 0$. As time increases, both areas strengthen and we see an increase in pressure magnitude, whilst

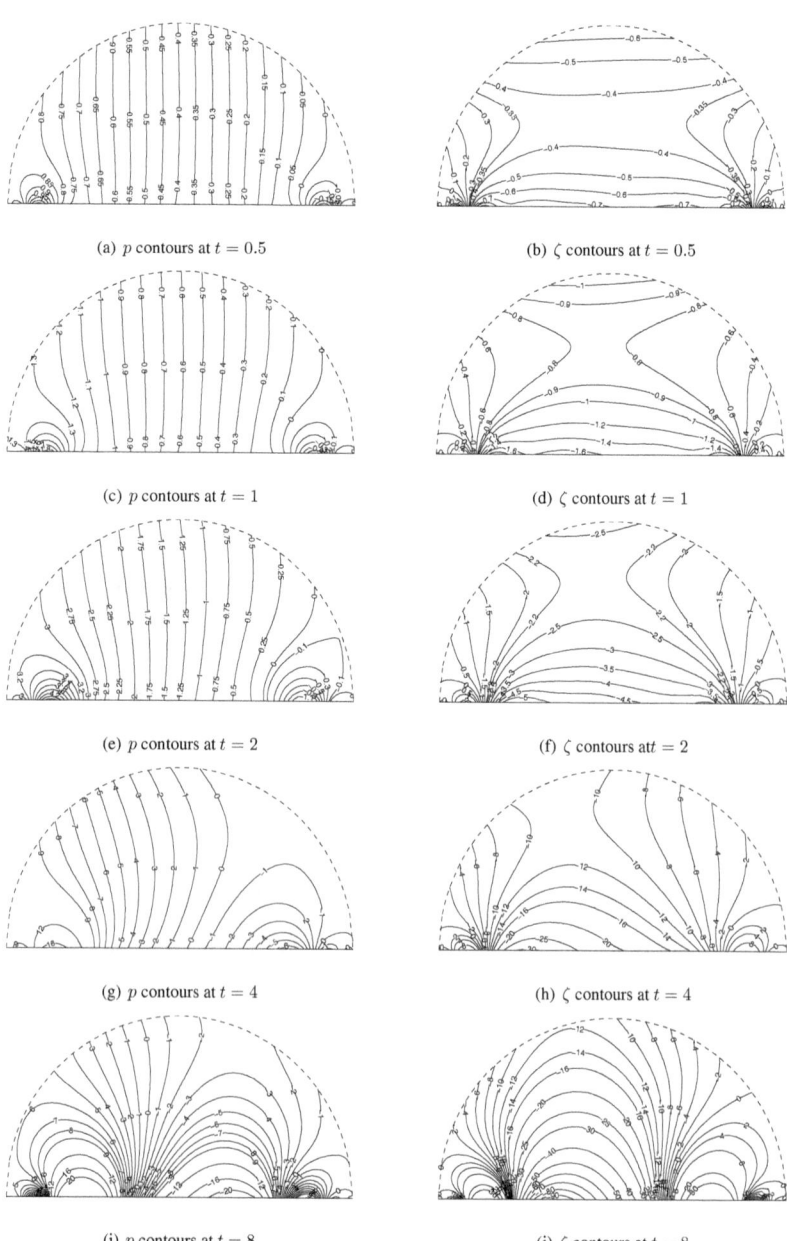

(a) $p$ contours at $t = 0.5$       (b) $\zeta$ contours at $t = 0.5$

(c) $p$ contours at $t = 1$       (d) $\zeta$ contours at $t = 1$

(e) $p$ contours at $t = 2$       (f) $\zeta$ contours at $t = 2$

(g) $p$ contours at $t = 4$       (h) $\zeta$ contours at $t = 4$

(i) $p$ contours at $t = 8$       (j) $\zeta$ contours at $t = 8$

Figure 4.1: Pressure and vorticity contours for $t = 0.5, 1, 2, 4, 8$ in the interacting semicircular droplet.

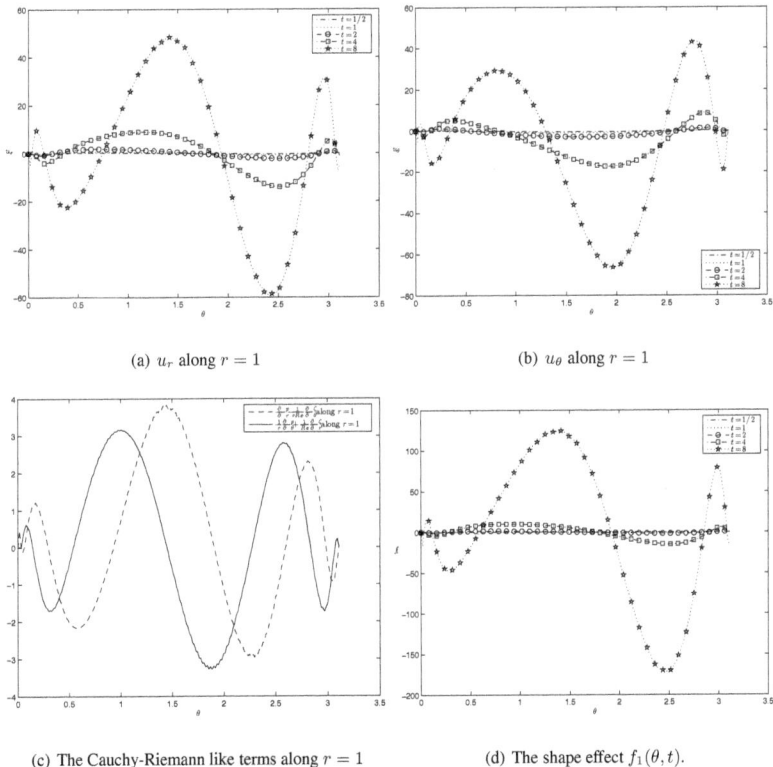

(a) $u_r$ along $r = 1$    (b) $u_\theta$ along $r = 1$

(c) The Cauchy-Riemann like terms along $r = 1$    (d) The shape effect $f_1(\theta, t)$.

Figure 4.2: Results of the interacting case for comparison with the idealised interface condition case.

elsewhere the pressure diffuses through the core of the droplet. At $t = 4$ we notice a third area

begin to increase in size near $\theta = 0$, as the low pressure region of earlier is pushed to the left

- similar to the results seen in the idealised interface conditions example. As time increases

further, the number of contour lines near $y = 0$ does also, suggesting a high pressure gradient

along the wall by $t = 8$.

The vorticity appears weak and well mixed in the early stages. Two areas corresponding

spatially to the high and low pressure regions are present near $\theta = 0, \pi$. As time increases

so too does the magnitude of vorticity and a pattern emerges similar to that seen in the ideal

interface example.

We notice also the difference between the images at $t = 4$, figures (4.1g), (4.1h), and those at

$t = 8$, figures (4.1i), (4.1j). While the system may well be steady in $p$, $\zeta$ by $t = 8$, it is certainly

not by $t = 4$. This is unlike our previous examples. Here, while $p$ and $\zeta$ remain functions of

time, $u_r$ and $u_\theta$ (governed by (3.0.11) and (3.0.12) respectively) will not yet have settled in to

a pattern of linear growth with time.

As we did in Chapter 2.2 we include plots of the interface values in figure (4.2) where curves are

drawn at $t = 1/2, 1, 2, 4, 8$. In the current problem, these figures serve an illustrative purpose

only: if $p$, $\zeta$ are not steady at $t = 4$, any comparison to the state at $t = 8$ will not be conclusive.

However, the images for $u_r$, figure (4.2a), and $u_\theta$, figure (4.2b), along $r = 1$ show that $\mathbf{u}$

increases by a little more than double over that time period. Comparison of the velocities to

the Cauchy-Riemann like terms displayed in figure (4.2c) show that the general shapes are the

same. A plot of $f_1$ in figure (4.2d) suggests the shape effect grows a little faster than $t^2$ - all as

one would expect with linear velocity growth in time in a system which has not yet reached a

steady state in $p$, $\zeta$.

It is figure (4.3) however which gives us confidence in suggesting that $u_r, u_\theta \sim t$ and $f_1 \sim t^2$. In (4.3a) a plot of $u_r$ and $u_\theta$ at the top of the droplet $(r, \theta) = (1, \pi/2)$ is plotted against time. The curve begins to settle by about $t = 4$ and becomes straight - showing linear velocity growth - between $t = 4$ and $t = 6$. Figure (4.3b) shows a plot of $f_1^{1/2}$ against $t$. The two terms begin to show a linear relationship at around $t = 2$ matching to those results seen in the idealised interface conditions example of the previous chapter.

For visualisation purposes, we also include a Cartesian plot of droplet movement for different values of $\epsilon$ in figure (4.4), similar to that seen in figure (3.25b) of the previous chapter. The movement appears realistic, with fluid moving into the area of low pressure which is offset to the right by shear forces.

## 4.2 The later temporal stage

Clearly, once $\epsilon f_1 = O(1)$, and the shape of the droplet is no longer a semicircle, the current results for flow in air over a semicircle will no longer hold. Finding a numerical solution to the flow over an obstacle with nonlinear shape for $Re = O(1)$ is a non trivial task and beyond the scope of this thesis.

We may however provide some brief analysis of the later stage: as $t \to 1/\epsilon^{1/2}$ in this case, where $f_1 \sim t^2$. To do so, we rescale $t$ as $t = \epsilon^{-1/2}T$, bearing in mind that the original water velocities were $O(\epsilon)$, and apply the following expansions to the nondimensionalised Navier-Stokes equations for the flow in water (1.2.10)

$$\hat{\mathbf{u}}_W = \epsilon^{1/2}\mathbf{U}_0 + \epsilon\mathbf{U}_1 + \cdots, \tag{4.2.1}$$

$$\hat{\zeta}_W = \epsilon^{1/2}Z_0 + \epsilon Z_1 + \cdots, \tag{4.2.2}$$

$$\hat{p}_W = \epsilon P_W + \cdots. \tag{4.2.3}$$

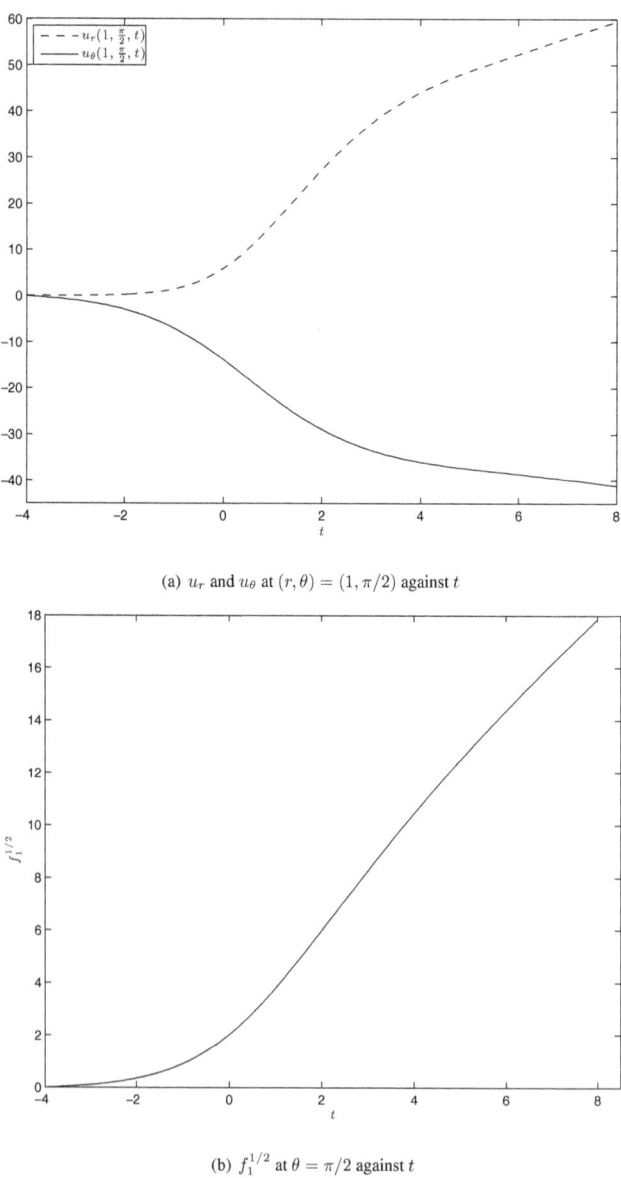

(a) $u_r$ and $u_\theta$ at $(r, \theta) = (1, \pi/2)$ against $t$

(b) $f_1^{1/2}$ at $\theta = \pi/2$ against $t$

Figure 4.3: Results to demonstrate linear velocity growth with time in the interacting model.

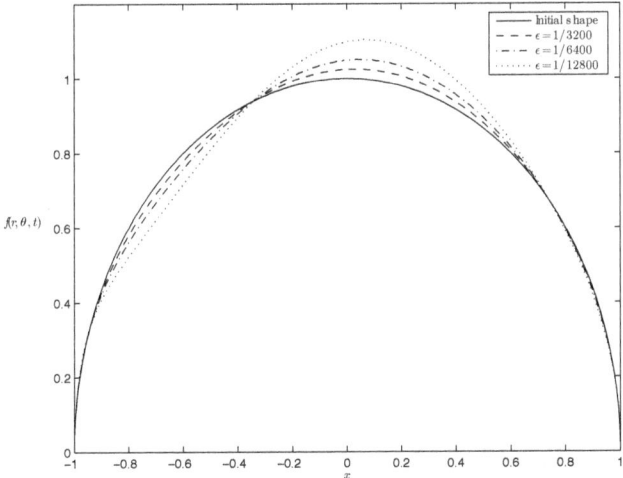

Figure 4.4: The shape effect in Cartesian geometry of the interacting semicircular droplet.

Doing so yields an $O(\epsilon^{1/2})$ balance of

$$\nabla^2 \mathbf{U}_0 = 0, \qquad (4.2.4)$$

so that to leading order, the velocities within the droplet are harmonic.

The scalings in air remain as before and a solution to the quasi-steady air flow over the nonlinear shape will provide the interface conditions:

$$\hat{\zeta}_W = \frac{\mu_A}{\mu_W} \zeta_A, \qquad (4.2.5)$$

$$p_W = p_A. \qquad (4.2.6)$$

Given that $\mu_A/\mu_W = O(\epsilon)$ (see Chapter 1 ) and $\zeta_A = O(1)$, (4.2.5) suggests we must also have $Z_0 = 0$.

Extending these results to higher order accuracy, an $O(\epsilon)$ balance of the Navier-Stokes equations sees the inertial terms re-enter the dynamics:

$$\frac{\partial \mathbf{U}_0}{\partial t} + (\mathbf{U}_0 \cdot \nabla) \mathbf{U}_0 = -\nabla P_W + \mathrm{Re} \left( -\frac{1}{r} \frac{\partial Z_1}{\partial \theta}, \frac{\partial Z_1}{\partial r} \right), \tag{4.2.7}$$

and the governing system takes the form of a forced Euler equation for $\mathbf{U}_0$, with the forcing term

written as a vector. It is relatively simple to show that (4.2.7) also implies that $Z_1$ is harmonic

$$\nabla^2 Z_1 = 0, \tag{4.2.8}$$

so that the forcing in (4.2.7), $-\frac{1}{r}\frac{\partial Z_1}{\partial \theta}$ radially and $\frac{\partial Z_1}{\partial r}$ azimuthally, can be determined numer-

ically using (4.2.5) and the Cauchy-Riemann equations for $Z_1$ and $P_W$ along $\theta = 0, \pi$.

The equations in this form leave the kinematic condition as

$$U_{0r} = \frac{\partial f}{\partial t} + U_{0\theta} \frac{\partial f}{\partial \theta}, \tag{4.2.9}$$

and we can begin to see how this would work numerically if a solution to the flow over an

obstacle of shape $f$ were known. Such an investigation is not attempted for this model.

## 4.3 Comparison to Smith and Purvis

One key question remains: for shape $f = f_0 + \epsilon f_1$, why does the circular seed have $f_1 \sim t$, as

predicted by Smith and Purvis [62], while the square droplet and the semicircular droplet both

have $f_1 \sim t^2$?

We suggest that the explanation lies in our decision to take the interfacial conditions as contin-

uous pressure and vorticity, as becomes clear if we consider the full stress conditions. In polar

form, after suitable non dimensionalisation, these require that tangentially:

$$\nu_A \left( \zeta_A - 2 \frac{\partial u_{\theta A}}{\partial r} \right) = \nu_W \left( \zeta_W - 2 \frac{\partial u_{\theta W}}{\partial r} \right), \tag{4.3.1}$$

and perpendicular to the interface:

$$-p_A + 2\nu_A \frac{\partial u_{rA}}{\partial r} = -p_W + 2\nu_W \frac{\partial u_{rW}}{\partial r}. \tag{4.3.2}$$

Our model is based on $\rho_A/\rho_W = \epsilon \ll 1$ and $\mu_A/\mu_W = O(\epsilon)$ (see chapter (1), suggesting that $\nu_A/\nu_W$ must be $O(1)$, but without placing any restrictions on the size of $\nu_A$, $\nu_W$ themselves.

We see now that taking $\nu_A$, $\nu_W \ll 1$ and rescaling by

$$\zeta_W = \frac{\widehat{\zeta}}{\nu_W}, \qquad \zeta_A = \frac{\widehat{\zeta}}{\nu_A} \qquad (4.3.3)$$

reduces the leading order terms of (4.3.1) and (4.3.2) to a continuous pressure and vorticity requirement across the interface. This is the case for which our model holds. However, without this additional restriction on $\nu_A, \nu_W$, we notice that as $t$ becomes large, any linear growth in $u_r$, $u_\theta$ (as seen in the flow within a semicircular droplet) will cause the $\dfrac{\partial}{\partial r}$ term to dominate the right hand side of the stress condition both tangentially (4.3.1) and perpendicular to the interface (4.3.2). This will be unbalanced by the terms on the left hand side - fixed as $O(1)$ by the flow in air - which forces the velocities within the droplet to become $O(1)$ also.

We postulate that had we applied the full stress conditions to our model, rather than just pressure and vorticity, we would not expect to see $f_1 \sim t^2$ or linear velocity growth in time, but instead have velocities which saturate as time increases. This directly implies that instead $f_1 = O(t)$ and the later temporal stage comes into operation as $t \to 1/\epsilon$ as predicted by Smith and Purvis [62].

In the case of the semicircular droplet, and its interaction with the surrounding air flow, it would be possible to test this hypothesis by repeating the simulation with the full stress conditions in place. However, given the validity restriction of the outer flow as $t \to \infty$ and the resolution restriction from the flow in air, we choose instead to focus on the special case mentioned earlier : the triple deck droplet. In that model which follows in Chapter 8, the full stress conditions may be applied and we do indeed see the $f_1 \sim t$ emerge.

Before doing so we present an additional investigation, arising from matters raised in the present model, but separate to droplet deformation. The so called dry-wet-dry problem concerns a no slip condition in a fluid governed by Stokes flow, equivalent to the problem associated with the Cauchy-Riemann equations for pressure and vorticity. We explore possible solutions to these equations, looking particularly at the region near a point of contact between solid, liquid and gas.

Chapter 5

# Stokes equations near a wall; solutions to the Cauchy-Riemann equations for pressure and vorticity

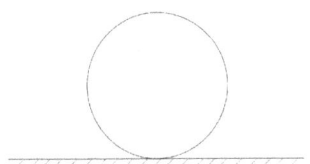

Figure 5.1: A droplet attached to a wall.

As we have seen earlier in this part of the thesis, along a wall where a no slip condition applies in fluid which is governed by the Stokes equations, pressure and vorticity require must satisfy the Cauchy-Riemann equations. A derivation of this may also be found in Pozrikidis [43]. Such a result allows us to study the 2D Stokes equations in the context of complex functions as in Langlois [30]. The solutions to these equations turn out to be an intriguing analytical problem, as explored by Mikhlin [36], with similarities to integro differential equations found in the theory of aircraft wings of finite span, Muskhelishvili [38] and the inhomogeneous Fredholm equations seen in signal processing or, for example, Fermo and Russo [17].

The behaviour of solutions for pressure and vorticity near a contact point between air, water and solid are particularly interesting. To investigate this relationship, we consider a model problem of droplet resting with a tiny part of its surface in contact with a solid wall, as shown

in figure (5.1). Local to the point of contact, this is equivalent to a liquid layer resting on a part

hydrophillic, part hydrophobic wall, so that the wall is wet on the hydrophillic section, and dry

on the hydrophobic section. For this reason, we name the model the dry-wet-dry problem.

## 5.1   The dry wet dry problem

Scaling the velocity and pressure within the droplet via the small ratios method $(\hat{\mathbf{u}}, \hat{p}) = \epsilon(\mathbf{u}, p)$

leads directly to the unsteady Stokes equations as we have seen before in Chapter 1. If we

assume the air film between the droplet and the hydrophobic section of the wall to be infinites-

imally thin, the system reduces to a one dimensional problem as shown in figure (5.2). Along

Figure 5.2: A graphical representation of the dry-wet-dry problem.

the hydrophillic section, between $x = -1$ and $x = 1$ the droplet is attached, thus the velocities

must satisfy the no-slip and no penetration conditions $u = v = 0$, and the Cauchy-Riemann

equations hold for pressure and vorticity. A direct result of which suggests the existence an

analytic function $F(z)$ with real part equivalent to the vorticity and imaginary part equal to the

pressure:

$$\frac{\partial p}{\partial x} = -\frac{\partial \zeta}{\partial y},$$ (5.1.1)

$$\frac{\partial p}{\partial y} = \frac{\partial \zeta}{\partial x},$$ (5.1.2)

with $F(z) = \zeta + ip$

where the Reynolds number is taken as unity throughout this chapter.

Since this analytic function $F(z)$ is defined in the upper half-plane, $\text{Im}\,[z] > 0$, a Cauchy-

Hilbert transform describes the relationship between the surface pressure and surface vorticity of the system. That is along $y = 0$, for vorticity $\zeta(x) = \operatorname{Re}[F(z)]$ the pressure will be given as $p(x) = \operatorname{Im}[F(z)]$ up to an additive constant:

$$p(x) = \frac{1}{\pi} \int_{-\infty}^{\infty} \frac{\zeta(s)}{x - s} \, ds, \qquad (5.1.3)$$

$$\zeta(x) = -\frac{1}{\pi} \int_{-\infty}^{\infty} \frac{p(s)}{x - s} \, ds. \qquad (5.1.4)$$

An overview of the transform may be found in Carrier [10] or Ablowitz [1].

It is worth noting that taking the $x$ derivative of the mixed boundary conditions, (5.1.1) and (5.1.2), suggest that $\zeta'(x)$ and $p'(x)$ also satisfy the Cauchy-Riemann equations, where prime denotes differentiation with respect to $x$. This would imply that

$$\tilde{F}(z) = \frac{\partial \zeta}{\partial x} + i \frac{\partial p}{\partial x}, \qquad (5.1.5)$$

is also analytic, and the form of the relationship between $p$ and $\zeta$ holds for $\zeta'(x)$ and $p'(x)$ also. As long as $\zeta'(x), p'(x) \to 0$ as $|x| \to \infty$, our solution to equations (5.1.3), (5.1.4) can only be out by an additive constant regardless of the pressure and vorticity contribution at infinity.

Our main interest lies in the section of the real axis in which the droplet is attached to the wall; $x \in (-1, 1)$. For such an investigation we take $\zeta$ and $p$ as known for $x \in (-\infty, -1]$ and $x \in [1, \infty)$. Considering the expression for vorticity, (5.1.4), it is possible to eliminate the unknown pressure by splitting the integral,

$$\zeta(x) = -\frac{1}{\pi} \left\{ \int_{-\infty}^{-1} + \int_{1}^{\infty} \right\} \frac{p(s)}{x - s} \, ds - \frac{1}{\pi} \int_{-1}^{1} \frac{p(s)}{x - s} \, ds, \qquad (5.1.6)$$

the first part of which is known. Some of the second part is also known in a sense from (5.1.3), and we indeed use (5.1.3) to split (5.1.6) again, giving

$$\zeta(x) = -\frac{1}{\pi} \left\{ \int_{-\infty}^{-1} + \int_{1}^{\infty} \right\} \frac{p(s)}{x - s} \, ds - \frac{1}{\pi^2} \int_{-1}^{1} \left\{ \int_{-\infty}^{-1} + \int_{1}^{\infty} \right\} \frac{\zeta(t)}{(s - t)(x - s)} \, dt \, ds$$

$$(5.1.7)$$

$$- \frac{1}{\pi^2} \int_{-1}^{1} \int_{-1}^{1} \frac{\zeta(t)}{(s - t)(x - s)} \, dt \, ds.$$

All of this, but for the last term, is known. It seems sensible to group the known part together

as $k(x)$ (later, in the numerical calculations we choose to take $k(x) = x^2$). Hence

$$\zeta(x) = k(x) - \frac{1}{\pi^2} \int_{-1}^{1} \zeta(t) \int_{-1}^{1} \frac{1}{(s-t)(x-s)} \, ds \, dt. \qquad (5.1.8)$$

Using partial fractions on the integral with respect to $ds$ in (5.1.8) yields an exact integration,

to give the full equation left to investigate in terms of vorticity only,

$$\zeta(x) = k(x) - \frac{1}{\pi^2} \int_{-1}^{1} \frac{\zeta(t)}{x-t} \ln \left| \frac{(1-t)}{(1+t)} \frac{(1+x)}{(1-x)} \right| \, dt. \qquad (5.1.9)$$

The next task then, is to solve this equation (5.1.9) for the unknown vorticity $\zeta(x)$ in $(-1, 1)$.

Since $\zeta$ is effectively expressed as a function of itself, it is convenient to define $x$ and $t$ on the

same grid when computing a numerical solution. Doing this does not cause difficulties at $x = t$

despite the inverse $(x - t)$ term, as the $\ln 1$ term dictates a zero contribution to the integral. At

$x = -1, 1$, and $t = -1, 1$ however, the logarithmic term becomes infinite. To deal with this in

a satisfactory way numerically, in particular when examining the behaviour $x \to 1$, we instead

consider the expression to be

$$\zeta(x) = k(x) - \frac{1}{\pi^2} \int_{-1+\hat{\epsilon}}^{1-\hat{\epsilon}} \frac{\zeta(t)}{x-t} \ln \left| \frac{(1-t)}{(1+t)} \frac{(1+x)}{(1-x)} \right| \, dt, \qquad (5.1.10)$$

where $x \in (-1 + \hat{\epsilon}, 1 - \hat{\epsilon})$ in the limit as $\hat{\epsilon} \to 0$.

It is simple to solve this using an iterative process for $\zeta$, by applying a discretization method,

identifying $i$ as the counter for $x$ and $j$ as the counter for $t$. We make an initial guess for $\zeta$ of

zero everywhere, and continue until successive iterations of the equation

$$\zeta_i = k_i - \frac{1}{\pi^2} \sum_j \frac{\zeta_j}{x_i - t_j} \ln \left| \frac{(1-t_j)}{(1+t_j)} \frac{(1+x_i)}{(1-x_i)} \right| \delta t, \qquad (5.1.11)$$

satisfy our convergence condition,

$$\frac{\max|\zeta_i - \zeta_i^{old}|}{|\text{ mean }(\zeta)|} < 10^{-5}, \qquad (5.1.12)$$

where *old* refers to the vorticity $\zeta$ at the previous iteration, and *mean* is the average value of

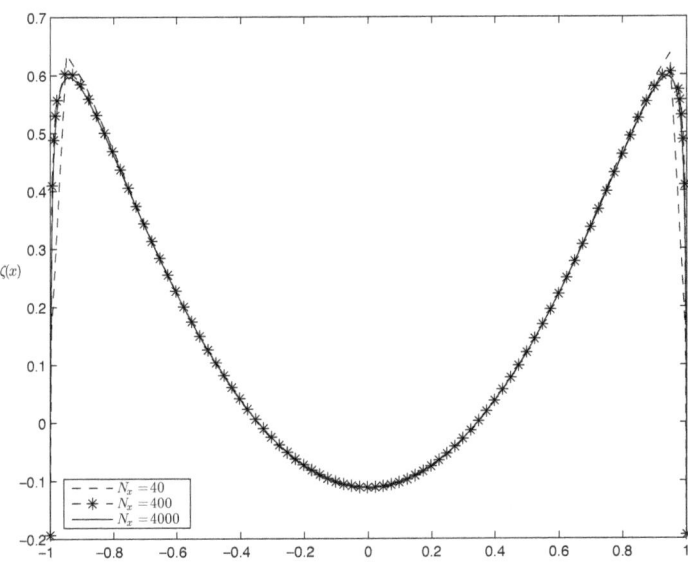

Figure 5.3: Plot of $\zeta$ against $x$ for $40, 400$ and $4000$ grid points.

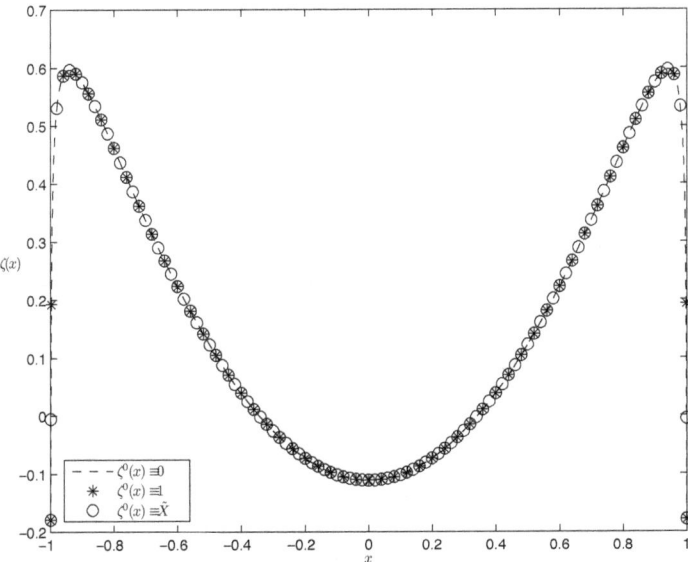

Figure 5.4: $\zeta$ found from 4000 grid points with initial guesses of $1, 0$ and a computer generated random number $\tilde{X}$

$\zeta$ across all discretized grid points. The right hand side of (5.1.11) is treated as known at each iteration of course, and then updated. This iterative process, in particular with no $\zeta$ contribution at the end points, means that we do not need to fix any boundary values.

The results for $\zeta$ against $x$ are given in figure (5.3). The plots are made with $\hat{\epsilon} = 5\text{x}10^{-4}$ and the grids have $N_x = 40$, $400$ and $4000$ points, so that

$$\delta x = \delta t = \frac{2 - 2\hat{\epsilon}}{N_x - 1}. \tag{5.1.13}$$

Also, to fix the problem, $k(x) = x^2$ is taken here. The numerical process converges after around 10 to 15 iterations, and the main point overall seems clear from the graph, namely that the solution settles in well with grid refinement. The vorticity is symmetrical in shape and dips below zero at $x = 0$, peaking near $x = -1$ and $x = 1$, before it appears to again become negative at the edges. This comes as a surprise, since the logarithmic term in (5.1.9) is infinite near $x = 1$. A good first guess would be to expect the vorticity to become zero at the edges, to avoid an infinite contribution from the log term and allow a balance in the equation but the numerical results do not seem to support this.

Repeating the procedure with a different initial guess seems to have little effect on the result. Plots are shown in figure (5.4) of a converged vorticity from initial guesses of zero, one, and a computer generated random number $\tilde{X}$ everywhere. We see from the graph that the results appear to be identical and, crucially, all drop below zero as the graph approaches $x = 1$ and $x = -1$. To investigate further, we switch our focus exclusively to what happens in the neighborhood of these two points.

## 5.2   Behaviour as $x \to -1$ and $x \to 1$

Since the system is symmetrical about $x = 0$, we focus on what happens near $x = 1$. The unknown value of the vorticity makes it perhaps a little difficult to examine the behaviour of the

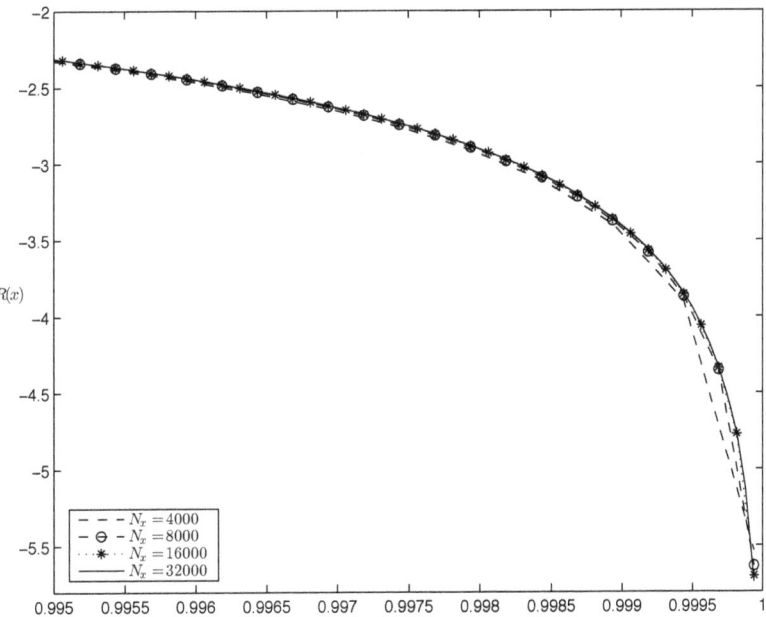

Figure 5.5: Close up plot of $R(x)$, defined in (5.2.1) with $\zeta = 1$ against $x$ for $4000, 8000, 16000$ and $32000$ grid points.

integral near this point, so it seems sensible to first consider the integral alone with the vorticity term set to one. For ease of notation, we refer to this integral as $R(x)$ from now on:

$$R(x) = -\frac{1}{\pi^2} \int_{-1}^{1} \frac{1}{x-t} \ln \left| \frac{(1-t)(1+x)}{(1+t)(1-x)} \right| dt. \tag{5.2.1}$$

Plots of $R$ close to $x = 1$, where $x, t \in (-1+\hat{\epsilon}, 1-\hat{\epsilon})$ and $\hat{\epsilon} = 6\text{x}10^{-5}$ can be seen in figure (5.5). Taking care near $x = t$, $R$ can be split into three parts, the last of which has an exact solution.

$$R(x) = -\frac{1}{\pi^2} \left\{ \text{p.v.} \int_{-1}^{1} \frac{\ln|1-t|}{x-t} dt - \text{p.v.} \int_{-1}^{1} \frac{\ln|1+t|}{x-t} dt + \ln \left| \frac{1+x}{1-x} \right| \text{p.v.} \int_{-1}^{1} \frac{1}{x-t} dt \right\}. \tag{5.2.2}$$

Examining each term individually in $R$ we can compare a numerical and analytical solution near $x = 1$, the latter found by considering each term as the real part of a complex problem.

## 5.3  The first term in $R$

First, address the numerical solution. Our motivation is to understand the bigger problem with a vorticity contribution where it is necessary to define $x$ and $t$ as equivalent, we do so here also, despite the singularity at $x = t$. At these points, we use Cauchy Principal Value, so that the first term in $R$ is found from the following,

$$-\frac{1}{\pi^2} \left\{ \int_{-1+\hat{\epsilon}}^{x-\delta x} \frac{\ln|1-t|}{x-t} dt + \int_{x+\delta x}^{1-\hat{\epsilon}} \frac{\ln|1-t|}{x-t} dt \right\}, \tag{5.3.1}$$

using $N_x = 4000, 8000, 16000, 32000$, with $\hat{\epsilon} = 6\text{x}10^{-5}$.

To find a local asymptote we take advantage of the similarities between our integral and the Cauchy-Hilbert transform by considering the following,

$$X = -\frac{1}{\pi} \text{p.v.} \int_{-1}^{1} \left( \frac{\ln|1-t|}{\pi} \right) \frac{1}{x-t} dt, \quad Y = \frac{\ln|1-x|}{\pi}, \tag{5.3.2}$$

and seek $G(z) = X + iY$ which is an analytic function of $z$. Given $Y$, $G(z)$ and $X$ will be unique, up to an additive constant. One form which works is

$$G(z) = \frac{1}{2\pi^2}(\ln z)^2, \tag{5.3.3}$$

$$= \left(\frac{(\ln r)^2}{2\pi^2} - \frac{1}{2}\right) + i\left(\frac{\ln r}{\pi}\right), \tag{5.3.4}$$

with $r$ defined as zero on $x = 1$. Suggesting that

$$-\frac{1}{\pi^2}\text{p.v.}\int_{-1}^{1}\frac{\ln|1-t|}{x-t}\,dt \sim \frac{(\ln|1-x|)^2}{2\pi^2} - \frac{1}{2} \quad \text{near } x = 1. \tag{5.3.5}$$

Plots of both solutions, numerical and analytic are shown in figure (5.6), where the solid line with black squares is the asymptote. The graph shows that both tend to infinity, and the analytic solution provides a good leading order approximation to the integral local to $x = 1$.

## 5.4 The last term in $R$

The final term in R, shown in (5.2.1), has an exact solution of

$$-\frac{1}{\pi^2}\left(\ln\left|\frac{1+x}{1-x}\right|\right)^2, \tag{5.4.1}$$

which allows us to test our numerical method of Section 5.1. Results are shown in figure (5.7) of the numerically calculated integral (found in the same way as the first in $R$), and the exact solution (solid line with black squares). The third term then has a negative and infinite contribution to $R$ as $x \to 1$ which does not cancel with the first term shown in figure (5.6).

## 5.5 An approximation to $R$

The second term of $R$ behaves as $\ln|1-x|$ near $x = 1$, so that with these $\ln|1-x|^2$ terms, we can neglect the second integral to leading order. Combining the log terms from the first and last integrals in $R$, we have

$$\frac{1}{2\pi^2}(\ln|1-x|)^2 - \frac{1}{\pi^2}\left(\ln\left|\frac{1+x}{1-x}\right|\right)^2 \sim -\frac{1}{2\pi^2}(\ln|1-x|)^2, \tag{5.5.1}$$

suggesting that we would expect, to leading order,

$$R \sim -\frac{1}{2\pi^2}(\ln|1-x|)^2 \quad \text{near } x = 1. \tag{5.5.2}$$

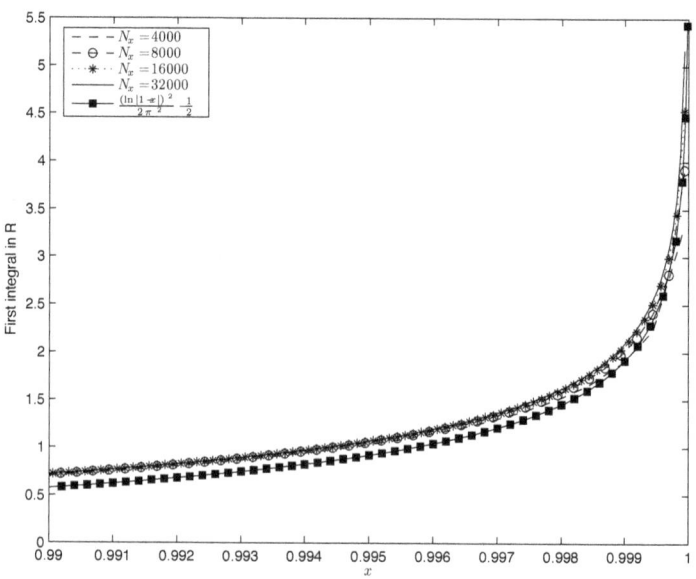

Figure 5.6: Plot of the numerically calculated first integral term in $R$ where $\zeta = 1$, against $x$ for finer grid sizes and its asymptote near $x = 1$.

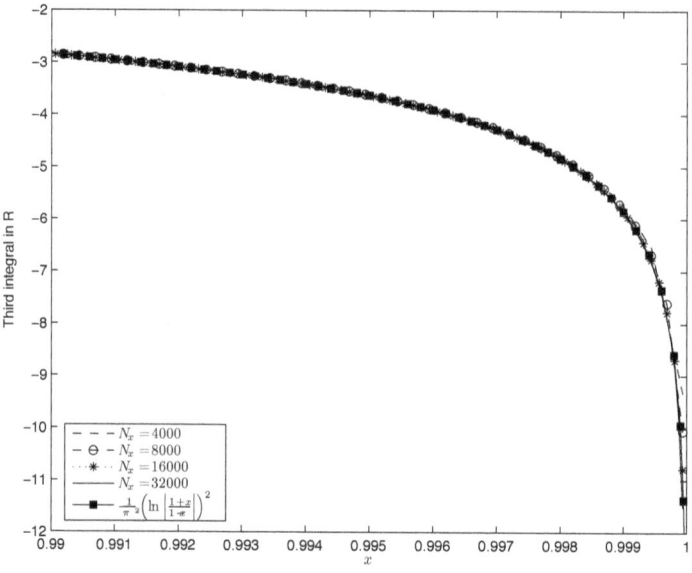

Figure 5.7: Plot of the numerically calculated third integral term in $R$ where $\zeta = 1$, against $x$ for finer grid sizes and its exact solution.

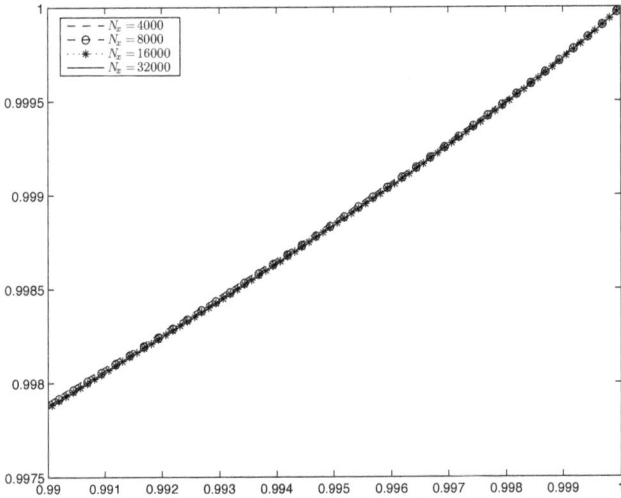

Figure 5.8: Plot of $1 - \exp\left[-\sqrt{-2\pi^2 R}\right]$ against $x$

For this to be the case, a plot of $x$ against

$$1 - \exp\left[-\sqrt{-2\pi^2 R}\right], \tag{5.5.3}$$

will give a straight line. (The negative square root is taken since $\ln|1 - x| < 0$). This plot is shown in figure (5.8). As the numerical computations do give straight lines near $x = 1$, we can have some confidence that $R$ does indeed take the form in (5.5.2).

## 5.6 Extension to the full problem

It remains to determine the value which $\zeta$ takes at $x = 1$. For ease of notation, we refer to the the integral term with a vorticity contribution as $S(x)$, as opposed to $R$ in (5.2.1) which was without the $\zeta$ term.

$$S(x) = -\frac{1}{\pi^2} \int_{-1}^{1} \frac{\zeta(t)}{x - t} \ln\left|\frac{(1 - t)(1 + x)}{(1 + t)(1 - x)}\right| \, dt, \tag{5.6.1}$$

making the full equation (5.1.9) simply $\zeta = k + S$. A plot of each of these three terms, found numerically in the same way as in the method in Section 5.1, close to $x = 1$ is shown in figure

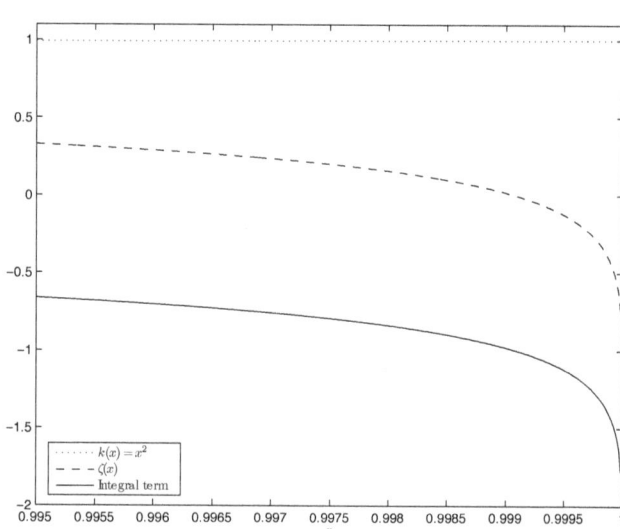

Figure 5.9: Plot of $\zeta$, in the case $k = x^2$, and the integral term $S(x)$ against $x$ for 128000 grid points. The governing equation is (5.1.9).

(5.9) for $N_x = 128000$ and $\hat{\epsilon} = 1.5 \times 10^{-5}$.

Our work of Sections 5.2 - 5.5, in the case where $\zeta = 1$ everywhere suggests that the numerics are fairly accurate and that we expect

$$S = O(\zeta(\ln r)^2). \tag{5.6.2}$$

However, the plot shows that even though $\zeta(x)$ is non zero near $x = 1$, $S(x)$ does not appear to behave logarithmically, and the equation balances well with

$$S = O(\zeta). \tag{5.6.3}$$

Putting (5.6.2) and (5.6.3) together, $S$ must take the form

$$S = A(\ln r)^2 + B \ln r + C + \cdots \tag{5.6.4}$$

$A, B$ and $C$ are constants dependant on the size of $\zeta$. For both (5.6.2) and (5.6.3) to hold,

$A = B = 0$.

This allows us to find a rough form for $\zeta$, as we can write down an expression for the leading order term in $S$ and solve it, knowing it must turn out to be zero;

$$S \sim \frac{-(\ln r)^2}{\pi} \int \frac{\zeta(t)}{x - t} \, dt. \tag{5.6.5}$$

Using the same technique as before, we take advantage of the similarities between this and the Cauchy-Hilbert transform by considering

$$X = \frac{-1}{\pi} \int \frac{\zeta(t)}{x - t} \, dt, \qquad Y = \zeta(t), \tag{5.6.6}$$

and seek the function $H(z) = X + iY$, which is analytic in $z$. Figure (5.9) suggests that $H(z) = z^n$, which would make $X$ and $Y$,

$$X = r^n \cos n\theta, \qquad Y = r^n \sin n\theta, \tag{5.6.7}$$

and $S$ and $\zeta$ on $\theta = \pi$,

$$S \sim r^n (\ln r)^2 \cos n\pi, \qquad \zeta \sim r^n \sin n\pi. \tag{5.6.8}$$

A value of $n = \pm 1/2$ would then make $A = 0$ in (5.6.4). Based on our system, we assume that other values of $n$ which give the leading order term in $S$ as zero, such as $n = \pm 3/2, \pm 5/2$, do not form an admissible part of the solution. Our numerics show $\zeta < 0$ near $x = 1$ suggesting that $n = -1/2$ and $\zeta \sim -r^{-1/2}$. To test this, figure (5.10) shows a plot of $1/\zeta^2$ for increasing grid refinement. The curve becomes straighter and appears to be possibly approaching the point $1/\zeta^2 = 0$ as $x \to 1$ which tends to support our argument, although the exact form of $\zeta$ may not be entirely clear.

The same type of analysis can be performed on the other leading terms in $S$, but we expect that $S$ will have an asymptote similar to $r^{-1/2}$ also. Figure (5.11) is a plot of $1/S(x)^2$

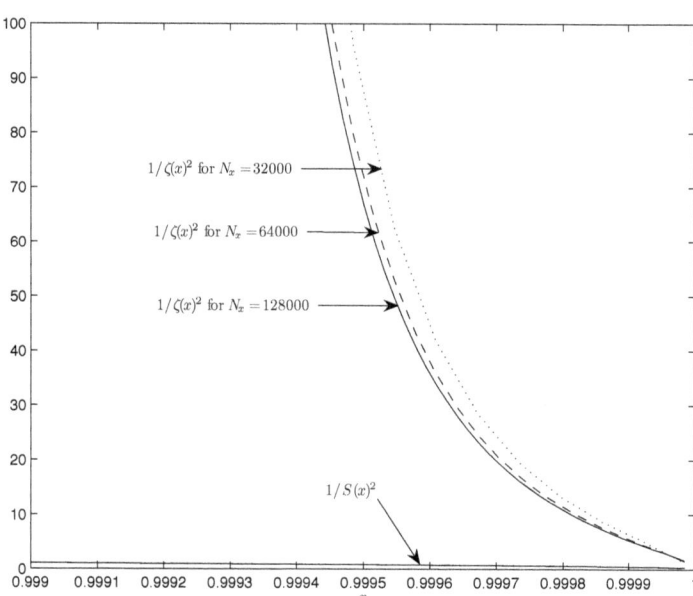

Figure 5.10: Plot of $1/\zeta^2$ against $x$. The governing equation is (5.1.9).

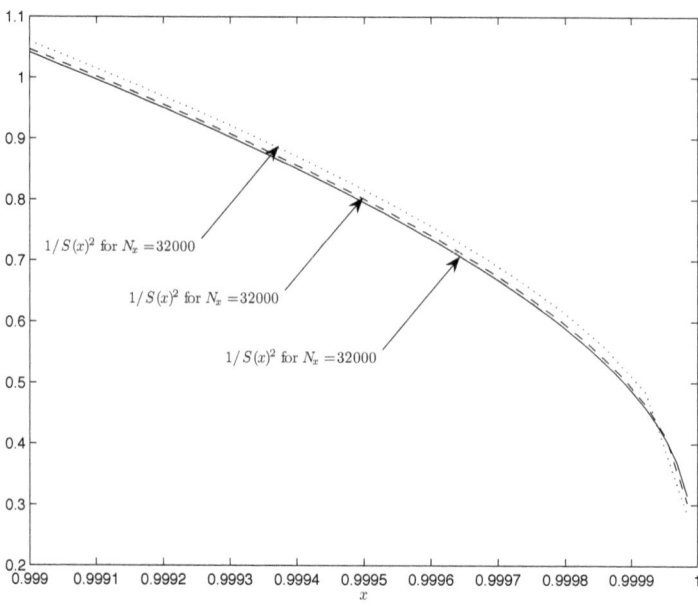

Figure 5.11: Plot of $1/S^2$ against $x$. The governing equation is (5.1.9).

against $x$ for increasing grid refinement. This is not inconsistent with the present suggestion of $n = -1/2$, accompanied by the logarithmic squared term in (5.6.8).

These results, if true are intriguing. Vorticity which behaves like $-r^{-1/2}$ near $x = 1 - r = 1$ suggests a small region near the solid-liquid-gas interface of clockwise particle rotation. More work is required to confirm these findings and to understand their effect on the 2D droplet, but the main feature of this chapter remains the one made earlier in figures (5.3) and (5.4), that the solution appears to tie together well for the present dry-wet-dry problem.

# Part II

# A droplet within a boundary layer

**Chapter 6**

# Boundary layer flow over a surface mounted

# obstacle

The investigation into droplet deformation which began in the previous part of the thesis has given us some insight into the dynamics of the two-fluid system, particularly in terms of the expected behavior at large times. However, the highly idealised nature of the models and boundary conditions makes it difficult perhaps to draw many firm conclusions, or to study the deformation of the droplet at large times.

In this part of the thesis we look at a practically important special case: the high Reynolds number flow of a surface mounted water droplet completely contained within the boundary layer of the external air flow. With this formulation it is possible to include the full interfacial stress conditions (to leading order) and to extend the time dependent simulation to the later temporal stage, where the interaction becomes nonlinear and the shape of the droplet becomes severely distorted. To our knowledge a model of this later temporal stage has not yet been attempted by other authors.

The general approach to modelling the deformation of the droplet follows that outlined of Part I but with one additional feature, as outlined in Chapter 1. Here, the solution for the external air flow is determined at each time step from the current interface shape. It is this feature which allows us to simulate the later temporal stage - in the previous part of the thesis the shape of the droplet did not feed back into the air flow solution.

To incorporate this two-way interaction of the fluids, we must obtain a general solution in air for flow over any hump, well or surface disturbance. The triple deck structure applied to a surface mounted obstacle contained within a boundary layer provides an opportunity to derive such an algorithm. That is the task of this chapter: to formulate the boundary layer flow in air over an obstacle and to present some example linear solutions.

A diagram of the flow regime is given in (6.1), where the obstacle pictured is solid (for now at least) and the air flow is subject to a no slip condition along its boundary. Since the second fluid, water, does not feature in this problem, we apply a more general approach by nondimensionalising the Navier-Stokes system (1.2.1) and (1.2.3) based on the characteristics of the flow in air (and not in the water as before) by using the scalings:

$$\mathbf{u}_A^* = U_A \mathbf{u}_A, \quad p_A^* = \rho_A U_A^2 p_A, \quad t^* = \frac{L}{U_A} t. \tag{6.0.1}$$

Here, the asterisk denotes the dimensional quantities, $\rho_A$ is the density of the air and $U_A, L$ are a characteristic velocity and lengthscale associated with the air flow. This serves to make the governing equations of this investigation as follows:

$$\frac{\partial \mathbf{u}_A}{\partial t} + (\mathbf{u}_A \cdot \nabla) \mathbf{u}_A = -\nabla p_A + \text{Re}_A^{-1} \nabla^2 \mathbf{u}_A, \tag{6.0.2}$$

$$\nabla \cdot \mathbf{u}_A = 0. \tag{6.0.3}$$

Later in Chapter 8, the solution to the triple deck flow over an obstacle which is presented in the following sections, coupled with the small density ratios assumption, will be used as the basis for the flow-in-air algorithm of our method of modelling droplet deformation.

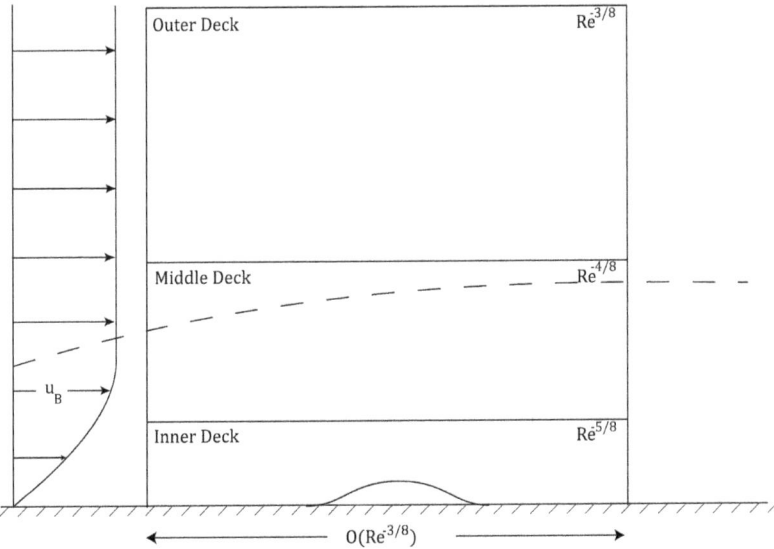

Figure 6.1: A schematic diagram of the triple deck flow over a surface mounted obstacle or droplet contained within a boundary layer.

In this more general form however the obstacle may be replaced by a second immiscible fluid and by adjusting the no slip condition along the boundary, a lubrication approximation may be applied to yield a Reynolds lubrication-type expression describing a steady two fluid interface as a function of $x$ only. This result does not depend on a small density ratio assumption. By investigating the families of solutions which this equation yields we may explore the effects of surface tension, gravity and shear stresses on the range of static droplet shapes and contact angles of a droplet contained in a triple deck flow. This is the focus of Chapter 7.

These two models, steady droplet shapes and unsteady droplet deformation, which follow later in the thesis are both based on the same problem in the air. That is our present task: to model the boundary layer flow over a surface mounted obstacle, like that in figure (6.1).

## 6.1   The triple deck structure

The wall upon which the obstacle is attached lies along the $x$ axis and $y$ is taken to be normal to the wall, we work in two spatial dimensions and the effects of compressibility are ignored throughout the flow geometry. The governing equations for the air flow are (6.0.2) and (6.0.3). Far from the wall we assume the flow to be inviscid and unidirectional and the Reynolds number $\mathrm{Re}_A$ is assumed to be large.

Close to the wall and upstream and downstream of the obstacle, an order of magnitude analysis may be applied in the standard way to (6.0.2) to reduce the characteristic of the momentum equation from elliptic to parabolic. The geometry of the flat plate also suggests a zero streamwise pressure gradient applies and thus the flow in air here is governed by the boundary layer equations of Prandtl [44]. In these regions, beyond the boundaries of the obstacle's influence, a Blasius similarity solution [8] may be used to satisfy the boundary layer equations, a good overview of which may be seen in White [74].

The obstacle itself has an influence on the boundary layer of lengthscale $O(\mathrm{Re}_A^{-3/8})$ shown first by Lighthill [32] both parallel and perpendicular to the plate. Within this region of influence, a viscous-inviscid interaction occurs causing a pressure perturbation and fluid displacement. The velocity profile becomes non linear and Prandtl's boundary layer equations no longer hold. To achieve a solution in the vicinity of the obstacle a triple deck structure must be applied.

The triple deck is based on the pressure-displacement mechanism described by Lighthill [32]. It was extended by Stewartson [66] and Messiter [35] who applied a three-tier structure in order to resolve the flow over the trailing edge of a flat plate as a remedy to the Goldstein near wake singularity [19]. Since then, the method has been used across many different problems where a disturbance within a boundary layer is large enough to cause a non linear velocity profile.

Flow over surface roughness, Smith and Burggraf [58], flow past corners, Rizzetta, Burggraf

and Jenson [50] and flow separation, Smith [55] are some such examples.

A diagram of the triple deck structure applied to our model is also shown in figure (6.1).

The region of flow which the obstacle influences is split into three subdomains or decks: an

inner viscous sublayer or inner deck, a middle deck in which viscous effects are secondary, and

an inviscid outer deck. The flow solutions to each deck are only valid within their respective

regions and must all merge with the two layer Blasius flow both upstream and downstream. The

decks are connected by the pressure perturbation and displacement function which Lighthill

described for a supersonic flow and interact with one another in the form of matched asymptotic

expansions.

As with Lighthill's work, the structure has a streamwise lengthscale of $x = \mathrm{Re}_A^{-3/8} X$ cen-

tered about the midpoint of the obstacle, while $y$ is rescaled accordingly within each deck.

Within the middle deck the transverse coordinate is rescaled as $y = \mathrm{Re}_A^{-1/2} \bar{Y}$. Assuming

that any change in velocity profile due to viscous effects is restricted to the inner deck, the mid-

dle deck sees the inertial terms balanced by the pressure gradient. Due to the viscosity in the

deck below, the middle deck does notice a change in the lateral displacement of the oncoming

fluid - known as the displacement function $A(X)$ - which it passes to the upper deck. If $u_B$ is

the oncoming Blasius flow to which the triple deck must match, a solution to (6.0.2) within the

middle deck will take the form:

$$u_A = u_B(\bar{Y}) + \mathrm{Re}_A^{-1/8} u_B'(\bar{Y}) A(X) + \cdots , \tag{6.1.1}$$

$$v_A = -\mathrm{Re}_A^{-1/4} A'(X) u_B(\bar{Y}) + \cdots , \tag{6.1.2}$$

$$p_A = \mathrm{Re}_A^{-1/4} \tilde{p}_A(X) + \cdots , \tag{6.1.3}$$

In the outer deck, where $y = \text{Re}_A^{-3/8}\hat{Y}$, the dynamics is controlled by the free-stream inter-

action and the flow must match to the unidirectional, inviscid flow above and the Blasius flow

up and downstream. Here, any fluid displacement caused by the obstacle is small and may be

linked to the deviation in pressure from the flat plate problem by a Hilbert integral. A solution

to (6.0.2) will take the form:

$$u_A = 1 + \text{Re}_A^{-1/4}\hat{u}_A(X,\hat{Y}) + \cdots, \tag{6.1.4}$$

$$v_A = -\text{Re}_A^{-1/4}\hat{v}_A(X,\hat{Y}) + \cdots, \tag{6.1.5}$$

$$p_A = \text{Re}_A^{-1/4}\hat{p}_A(X,\hat{Y}) + \cdots, \qquad\qquad \hat{p}_A(X,0) = \tilde{p}_A(X) \tag{6.1.6}$$

while matching to the middle deck yields the pressure-displacement law:

$$\tilde{p}_A(X) = \frac{1}{\pi}\int_{-\infty}^{\infty}\frac{A'(s)}{X-s}\,\text{d}s, \tag{6.1.7}$$

where the integral is interpreted as Cauchy principle value. Finally, within the inner deck, where

$y = \text{Re}_A^{-5/8}Y$ a boundary layer type disturbance occurs and a solution to (6.0.2) will take the

form:

$$u_A = \text{Re}_A^{-1/8}\tilde{u}_A(X,Y) + \cdots, \tag{6.1.8}$$

$$v_A = \text{Re}_A^{-3/8}\tilde{v}_A(X,Y) + \cdots, \tag{6.1.9}$$

$$p_A = \text{Re}_A^{-1/4}\tilde{p}_A(X) + \cdots, \tag{6.1.10}$$

Application of (6.1.8) to (6.1.10) to the momentum and continuity equations (6.0.2) and (6.0.3)

yields the inner deck equations:

$$\tilde{u}_A\frac{\partial \tilde{u}_A}{\partial X} + \tilde{v}_A\frac{\partial \tilde{u}_A}{\partial Y} = -\frac{\text{d}\tilde{p}_A}{\text{d}X} + \frac{\partial^2 \tilde{u}_A}{\partial Y^2}, \tag{6.1.11}$$

$$\frac{\partial \tilde{u}_A}{\partial X} + \frac{\partial \tilde{v}_A}{\partial Y} = 0. \tag{6.1.12}$$

These will be subject to a no slip condition along $Y = f(X)$, and must match to the middle deck above and the Blasius flow both upstream and downstream:

$$\tilde{u}_A = \tilde{v}_A = 0 \text{ along } Y = f(x), \tag{6.1.13}$$

$$\tilde{u}_A \to Y \text{ as } X \to \pm\infty, \tag{6.1.14}$$

$$\tilde{u}_A \sim \lambda(Y + A(x)) \text{ as } Y \to \infty, \tag{6.1.15}$$

Here, $\lambda$ is the value of $\partial u_B / \partial \bar{Y}$ at $\bar{Y} = 0$. With suitable transformation, details of which we did not think it necessary to include, it is sufficient to take the value of the constant $\lambda$ as unity.

The pressure-displacement law also applies here so that it remains to solve (6.1.11) and (6.1.12) in conjunction with (6.1.13), (6.1.14) (6.1.15) (with $\lambda = 1$ ) and (6.1.7) to obtain a complete description of the flow field.

## 6.2  Linearised triple deck flow over a solid obstacle

To solve the system of inner deck equations and boundary conditions outlined above in the case of flow over a surface mounted solid shape, we begin by mapping the $X$ axis in (6.1.11)-(6.1.15) to the surface of the obstacle using a Prandtl transposition or shift:

$$(X, Y, \tilde{u}_A, \tilde{v}_A, \tilde{p}_A) \to \left(x, y + f(x), u_A, v_A + u_A f'(x), p_A\right). \tag{6.2.1}$$

The symbols given in (6.2.1) for the transposed variables are chosen for easier notation and should not be confused with those of the previous section. Here, prime denotes differentiation with respect to $x$. Application of this shift leaves the governing equations (6.1.11)-(6.1.15) unchanged, aside from the wall condition which becomes $u_A = v_A = 0$ at $y = 0$ and the condition as $y$ tends to infinity (6.1.15) which now includes the hump shape,

$$u_A \sim y + f(x) + A(x). \tag{6.2.2}$$

We notice that if $f(x) = 0$ an exact solution of $u_A = y$, $p_A = v_A = A = 0$ exists. We may use this result as the basis of a perturbation by considering a hump shape of $f(x) = \delta F(x)$ in the limit as $\delta \to 0$. An approximate solution may be found using this approach with $\delta$ quantifying the deviation from the flat plate flow. In such a case, the velocities, pressure and displacement function will take the form of a series expansion in $\delta$:

$$u_A = y + \delta U_A + \cdots, \tag{6.2.3}$$

$$v_A = \delta V_A + \cdots, \tag{6.2.4}$$

$$p_A = \delta P_A + \cdots, \tag{6.2.5}$$

$$A = \delta\mathcal{A} + \cdots. \tag{6.2.6}$$

Later, in Chapter 8 where a nonlinear solution to the inner deck equations is sought, we remove the dependence of the air flow on the size of the obstacle, in effect, by taking $\delta = 1$. For the current purpose however, an $O(\delta)$ balance in the equations and boundary conditions yields a linearised system:

$$y\frac{\partial U_A}{\partial x} + V_A = -\frac{\mathrm{d}P_A}{\mathrm{d}x} + \frac{\partial^2 U_A}{\partial y^2}, \tag{6.2.7}$$

$$\frac{\partial U_A}{\partial x} + \frac{\partial V_A}{\partial y} = 0, \tag{6.2.8}$$

$$U_A = V_A = 0 \text{ on } y = 0, \tag{6.2.9}$$

$$U_A \to 0 \text{ as } x \to \pm\infty, \tag{6.2.10}$$

$$U_A \to \mathcal{A}(x) + F(x) \text{ as } y \to \infty, \tag{6.2.11}$$

while the form of the pressure-displacement relationship remains unchanged:

$$P_A(X) = \frac{1}{\pi} \int_{-\infty}^{\infty} \frac{\mathcal{A}'(s)}{X - s}\,\mathrm{d}s, \tag{6.2.12}$$

This formulation, (6.2.7) to (6.2.12), has been used widely as a standard approach to obtain solutions of the triple deck, Smith and Merkhin [60], Stewartson [67] and Jobe and Burggraf [26] being some such examples.

The advantage of this formulation becomes clear after application of the Fourier transform, the forward and inverse transforms of which we define as

$$\widehat{U}_A(k, y) = \int_{-\infty}^{\infty} U_A(x, y) e^{-ikx} \, dx, \qquad (6.2.13)$$

$$U_A(x, y) = \frac{1}{2\pi} \int_{-\infty}^{\infty} \widehat{U}_A(k, y) e^{ikx} \, dk, \qquad (6.2.14)$$

respectively, where a hat denotes Fourier Transformed variables. By taking the Fourier transform of the $y$ derivative of (6.2.7) then, it proves convenient to define and apply a new variable $\xi = (ik)^{1/3} y$, the result of which is to transform (6.2.7) into Airy's equation for $\dfrac{\partial \widehat{U}_A}{\partial \xi}$:

$$\frac{\partial^3 \widehat{U}_A}{\partial \xi^3} = \xi \frac{\partial \widehat{U}_A}{\partial \xi}, \qquad (6.2.15)$$

The solutions of Airy's equations are commonly taken to be of the form

$$\frac{\partial \widehat{U}_A}{\partial \xi} = B(k) Ai(\xi) + C(k) Bi(\xi), \qquad (6.2.16)$$

for Airy functions $Ai$ and $Bi$. In this case, $\dfrac{\partial \widehat{U}_A}{\partial \xi}$ is analogous to the perturbation of shear stress, the magnitude of which we expect to decrease as we move away from the surface of the hump. Since $\xi$ is complex and $Bi(\xi)$ increases with $|\xi|$, $C(k)$ must be identically zero.


Integrating (6.2.16) with $C(k) = 0$ gives an expression for the Fourier transformed velocity

$$\widehat{U}_A = B(k) \int_0^{\xi} Ai(s) \, ds, \qquad (6.2.17)$$

where the no slip condition (6.2.9) leads to a zero constant of integration. It is possible to eliminate the unknown function $B(k)$ by applying the Fourier transformed versions of the as yet unused boundary conditions: first, the integral of the Airy function $Ai$ from zero to infinity is equal to one third which couples with the boundary condition (6.2.11) to give

$$\widehat{A}(k) + \widehat{F}(k) = \frac{B(k)}{3}; \qquad (6.2.18)$$

second, the Fourier transform of the pressure displacement law (6.2.12) implies that

$$\widehat{P} = |k| \widehat{A}; \qquad (6.2.19)$$

and finally, the Fourier transform of the momentum equation along $y = 0$ (6.2.7) equated with

the derivative in $\xi$ of (6.2.16) gives

$$B(k) = \frac{(ik)^{1/3}\widehat{P}}{Ai'(0)},$$ (6.2.20)

where $Ai'(0)$ has a known value of $Ai'(0) = -\dfrac{1}{3^{1/3}\Gamma(\frac{1}{3})}$ for gamma function $\Gamma$ [2].

With these three, (6.2.18), (6.2.19) and (6.2.20), it is relatively simple to eliminate $B(k)$

and derive expressions linking the hump shape to the perturbed pressure, displacement and

velocity:

$$\widehat{P}_A(k) = \frac{3|k|Ai'(0)}{|k|(ik)^{1/3} - 3Ai'(0)}\widehat{F}(k),$$ (6.2.21)

$$\widehat{A}(k) = \frac{3Ai'(0)}{|k|(ik)^{1/3} - 3Ai'(0)}\widehat{F}(k),$$ (6.2.22)

$$\widehat{U}_A(k,\xi) = \frac{3|k|(ik)^{1/3}}{|k|(ik)^{1/3} - 3Ai'(0)}\widehat{F}(k)\int_0^{\xi} Ai(s)\,\mathrm{d}s.$$ (6.2.23)

This form of solution (6.2.21) to (6.2.23) suggests that the deviation from the flat plate flow

is characterised exclusively by the shape of the obstacle which the fluid passes over - as one

would expect.

It is fairly straightforward to investigate these expressions numerically, although an infinite

$x$ range is computationally expensive to achieve. Given that the boundary conditions (6.2.10)

already dictate that the flow must merge with the Blasius flow upstream and downstream, ap-

plying a simplifying assumption of periodic flow seems reasonable. Provided the period $2L$ is

chosen as sufficiently large, there will be no upstream influence from other obstacles.

We look then to use a discrete Fourier series in place in place of (6.2.13), although of course, in the limit $L \to \infty$, the Fourier series and Fourier transform are equivalent.

We have an advantage in deriving a discrete form of (6.2.13) as we know that the inverse transform of (6.2.21) to (6.2.23) must yield a real result: $P_A, \mathcal{A}, U_A$ all correspond to physical quantities. If $U_A$ is real, it must be that

$$\widehat{U}_A(-k, \xi) = \int_{-\infty}^{\infty} U_A(x, \xi) e^{ikx} \, dx = \text{c.c} \left[ \widehat{U}_A(k, \xi) \right], \tag{6.2.24}$$

where here, c.c denotes the complex conjugate. Likewise, the reverse is true. Any real quantity $g$ say, will have a transformed variable which satisfies

$$\text{c.c} \left[ \widehat{g}(k, \xi) \right] = \widehat{g}(-k, \xi). \tag{6.2.25}$$

This holds for $\widehat{F}$, since $F$ is real, and is also true for $|k|$, $(ik)^{1/3}$ and all other terms in the quotient of (6.2.21). Thus, the form of (6.2.21) will yield a real solution for $P_A$ as required. A similar argument holds for $\mathcal{A}$, although the equivalent for $U_A$ is a little trickier due to the integral of the Airy function.

It is this condition (6.2.25) which allows us to derive a shortcut. By manipulating the inverse form of the Fourier transform (6.2.14):

$$U_A(x, \xi) = \frac{1}{2\pi} \left[ \int_{-\infty}^{0} \widehat{U}_A(k, \xi) e^{ikx} \, dk + \int_{0}^{\infty} \widehat{U}_A(k, \xi) e^{ikx} \, dk \right], \tag{6.2.26}$$

$$= \frac{1}{2\pi} \left[ -\int_{\infty}^{0} \widehat{U}_A(-\hat{k}, \xi) e^{-i\hat{k}x} \, d\hat{k} + \int_{0}^{\infty} \widehat{U}_A(k, \xi) e^{ikx} \, dk \right], \tag{6.2.27}$$

where $k = -\hat{k}$, we may apply (6.2.25)

$$U_A(x, \xi) = \frac{1}{2\pi} \left[ \int_{0}^{\infty} \text{c.c} \left[ \widehat{U}_A(\hat{k}, \xi) \right] e^{-i\hat{k}x} \, d\hat{k} + \int_{0}^{\infty} \widehat{U}_A(k, \xi) e^{ikx} \, dk \right], \tag{6.2.28}$$

$$= \frac{1}{2\pi} \int_{0}^{\infty} \left[ \widehat{U}_A(k, \xi) e^{ikx} + \text{c.c} \left[ \widehat{U}_A(\hat{k}, \xi) \right] (k, \xi) e^{-ikx} \right] \, dk, \tag{6.2.29}$$

$$= \frac{1}{\pi} \int_{0}^{\infty} \mathcal{R}e \left[ \widehat{U}_A(k, \xi) e^{ikx} \right] \, dk. \tag{6.2.30}$$

By putting $k = \dfrac{\pi n}{L}$ for $n \in \mathbb{Z}$, $n \le N_k$ into this expression (6.2.30) and taking $\mathrm{d}n = 1$,

$N_k >> 1$, the integral may be replaced with a discrete sum over $n$:

$$U_A(x, \xi) = \frac{1}{L} \sum_{0}^{N} \mathcal{R}e \left[ \widehat{U}_A(k_n, \xi) \exp \left[ \frac{in\pi x}{L} \right] \right]. \tag{6.2.31}$$

Notice that although the sum is from 0 to $N_k$, the derivation means that all negative values of $n$

are also considered here - reducing computational time compared to a standard Fourier series.

This expression (6.2.31) replaces (6.2.14) as the inverse Fourier transform in any calculations,

while (6.2.13) remains as the forward transform.


To apply this scheme to any obstacle is fairly straightforward. With a known interface $F(x)$,

the relevant form of forward Fourier transform (6.2.13) can be applied to determine $\widehat{F}$. For each

integer value $n$, $k = \dfrac{n\pi}{L}$ is used in (6.2.21) and (6.2.23) and all results are summed together

using (6.2.31) to find the real space pressure and velocities.

### 6.2.1 Results

We have looked at two examples. First, the flow over a Witch of Agnesi shaped obstacle in

figure (6.2). We also present for comparison results from a similar computation featured in

a paper by Smith, Brighton, Jackson and Hunt [57] in figure (6.3). In both simulations the

obstacle shape has equation

$$F = \frac{1}{1 + x^2}, \tag{6.2.32}$$

which is particularly favorable for this type of calculation as the lack of corners reduces the

number of Fourier modes we need to consider to achieve reliable results. In both figures, (6.2)

and (6.3), the top image is a plot of the perturbation to wall shear, $\dfrac{\partial U_A}{\partial y}$ in our case, against the

triple deck scaled $x$. The middle image on both pages displays results for the perturbation to

pressure along the solid boundary, in our case $P_A$, against $x$, and finally the last image on each

page shows the shape of the obstacle considered, plotted on the same $x$ scale as those figures

above.

Our results are calculated on a grid with spacing $\delta x = 10^{-3}$ and $\delta y = \dfrac{\delta x^{1/2}}{4}$. The Fourier modes are summed from $n = 0$ to $N_k = 256$ and the period $2L$ is varied from $2L = 4\pi$ to $2L = 32\pi$. We notice in both the images for the pressure and shear, for periods longer than $L = 4\pi$ there is little difference in the solution, while even the shortest period $L = 2\pi$ appears to capture the overall trend well.

Comparison of our results to those of Smith et al. in figure (6.3) show close agreement overall. In the images for the pressure perturbations, we see the area of lowest pressure is near $x = 0$ but offset to the right from the center by the oncoming shear. The pressure does not quite return to ambient levels by $x = 4$, although this is more pronounced in Smith et al.'s example.

The trends appear to match for the shear stress perturbations too. Both sets peak to the left of $x = 0$ before becoming negative. Our results show a steep shear gradient after the minimum which is a feature not mirrored by Smith et al.

Our second example included is that of a parabolic hump. Results may be seen in figure (6.4). We also include results for comparison from another of Smith's papers [54] in figure (6.5)

Here the obstacle has shape described by

$$F = \max\left\{\frac{x(1 - \theta x)}{\theta}, 0\right\}, \tag{6.2.33}$$

where $\theta = (-3Ai'(0))^{3/4}$ as it is in [54]. Our results are again calculated on a grid of $\delta x = 10^{-3}$ and $\delta y = \dfrac{\delta x^{1/2}}{4}$, although on this occasion, $L$ is fixed at $L = 6\pi$ and $N_k$ is varied from 16 to 512. To match the order of Smith's results, in this example the images in (6.4) and (6.5) show (from top to bottom) perturbation to pressure along the wall of the obstacle, perturbation to wall shear, and the obstacle shape, all against $x$.

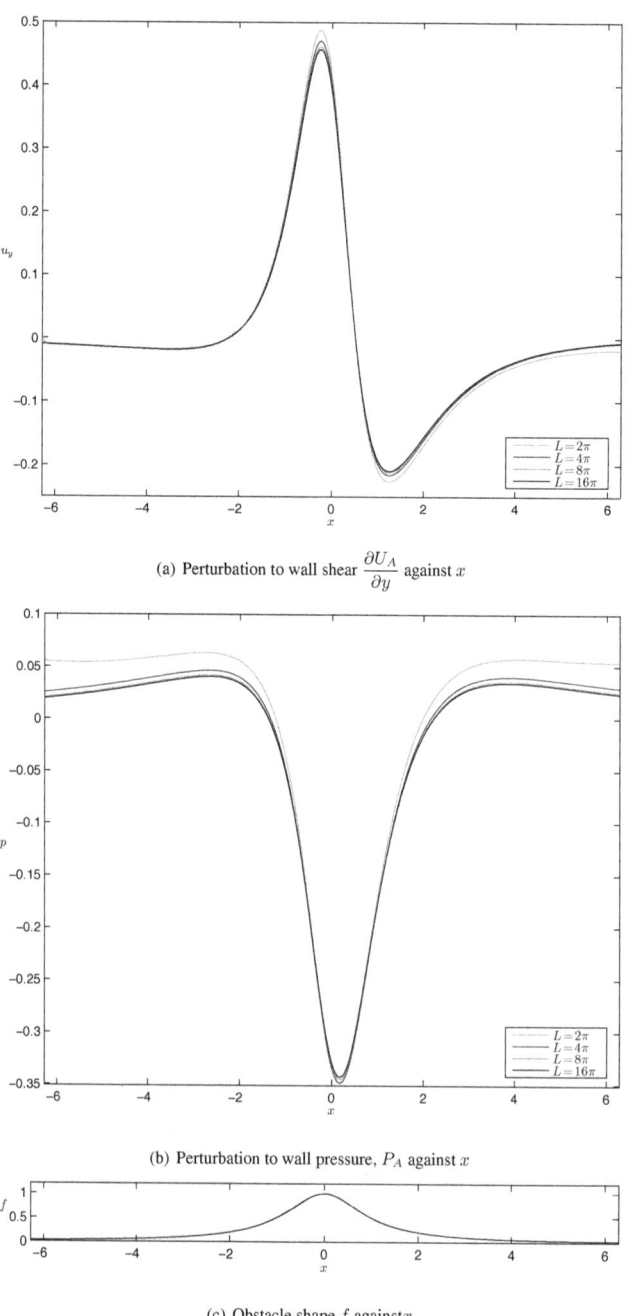

(a) Perturbation to wall shear $\dfrac{\partial U_A}{\partial y}$ against $x$

(b) Perturbation to wall pressure, $P_A$ against $x$

(c) Obstacle shape $f$ against $x$

Figure 6.2: Pressure and wall shear results for boundary layer flow over a Witch of Agnesi shaped obstacle

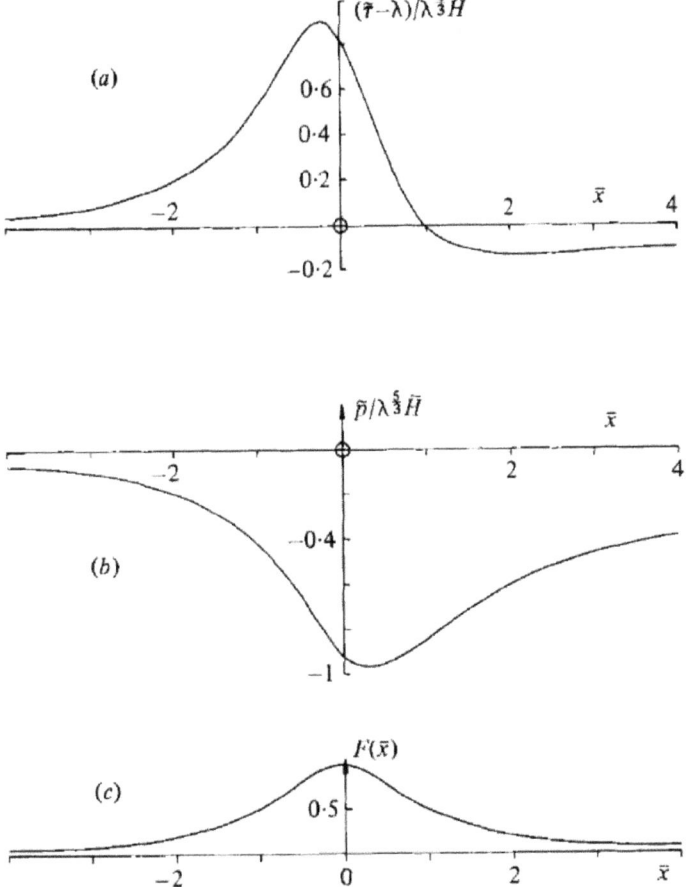

Figure 6.3: Results shown in [57] for the triple deck flow over a Witch of Agnesi hump. (a) is the perturbation to wall shear, (b) the perturbation to pressure and (c) the hump shape.

The corners on the parabolic hump mean much higher Fourier modes must be considered to capture the solution. This can be seen in figure (6.4) where even by $N_k = 64$, the solution has not yet settled. By $N_k = 512$ however, the sharp spike in pressure on the leading edge of the obstacle seen in (6.4a) has good definition and matches that seen in the dashed line of the Smith figure (6.5a). As in the previous example, the minimum pressure occurs to the right of the center of the obstacle and then slowly increases to just below ambient pressure.

The spikes in the plot for wall shear are well defined by $N_k = 512$ also. Again, the maximum occurs to the left of the center of the obstacle and we see here as in our results of the previous example a pronounced minimum shear point followed by a steep positive gradient on the trailing edge of the obstacle, matching Smith's results (6.5b) on this occasion.

Overall, the solutions behave sensibly with increased period and Fourier modes and we conclude that the method captures the solution well.

In Chapter 8 we apply the small-density-ratios assumption to the result shown here and extend the solution to include the nonlinear terms. For now however, we pause briefly on the overall aim of the thesis of modelling unsteady droplet deformation and use (6.2.7) - (6.2.12) to explore the steady shapes which a droplet can take in the presence of shear, gravity and surface tension.

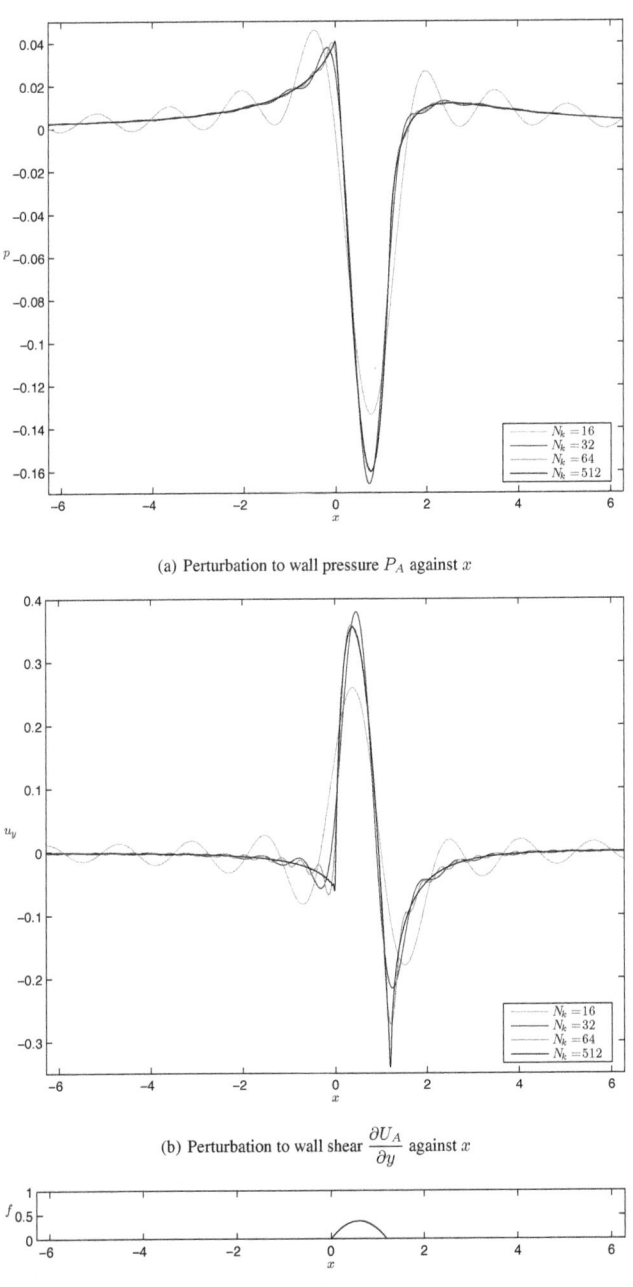

(a) Perturbation to wall pressure $P_A$ against $x$

(b) Perturbation to wall shear $\dfrac{\partial U_A}{\partial y}$ against $x$

(c) Obstacle shape $f$ against $x$

Figure 6.4: Pressure and wall shear results for boundary layer flow over a parabolic shaped obstacle

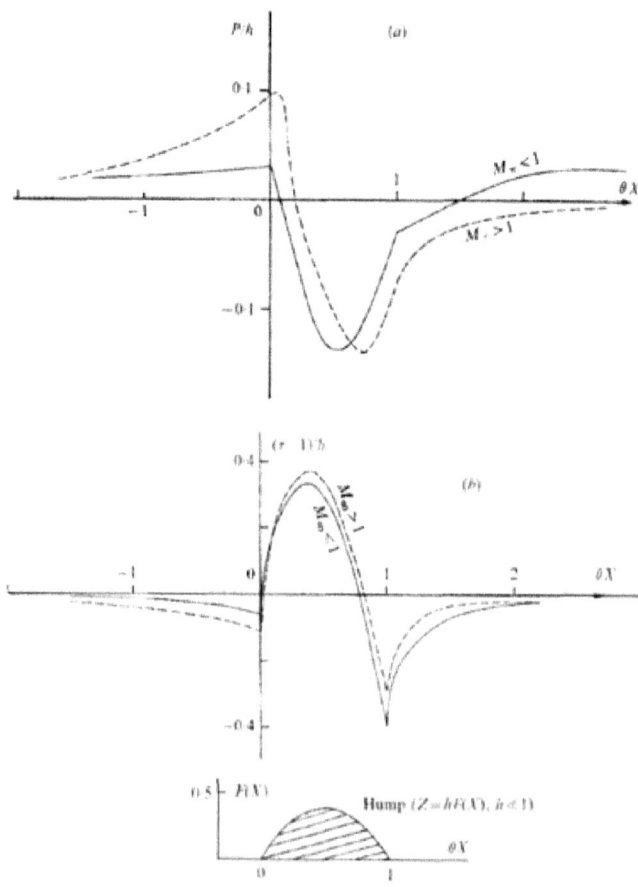

Figure 6.5: Results shown in [54] for the triple deck flow over a parabolic hump. (a) the perturbation to pressure, (b) is the perturbation to wall shear and (c) the hump shape.

## Chapter 7

# The effect of surface tension, gravity and shear stresses on a droplet within a boundary layer

In the previous chapter we outlined a linear solution for flow over a shallow surface mounted obstacle. Replacing this obstacle with a second fluid presents an opportunity to study the steady shapes of an attached droplet within a boundary layer. This side investigation is the focus of this chapter, where we adjust the solution of Chapter 6 to account for the second fluid, apply a lubrication approximation to the droplet and derive a Reynolds lubrication type expression, relating the shape of the two fluid interface to gravity, surface tension and shear forces.

A diagram of the flow regime is given in figure (6.1). In this case, the hump within the triple deck structure is a droplet, formed from a second immiscible fluid. For ease, we shall continue to refer to the second fluid as water although it should be noted that, unlike all other two fluid systems in the thesis, the results of this chapter do not rely on a small density ratio between the two fluids. To distinguish this example, we rename the interface of the two fluids here as $h(x)$.

To allow for interaction between the two fluids, the flow in water is nondimensionalised based on the characteristics of the flow in air, in a reverse process to that seen in Chapter 1, details of which we do not think it necessary to include. The droplet, like the flow in air, is

subject to the inner deck scalings of Chapter 6 and considered steady. Prandtl's transposition is

not applied and so we name the transverse coordinate here as y; related to that of the previous

chapter by $y = \mathrm{y} + h(x)$. Before any assumptions are made on the depth of the droplet then,

the dimensionless triple deck flow in water is governed by:

$$u_W \frac{\partial u_W}{\partial x} + v_W \frac{\partial u_W}{\partial \mathrm{y}} = -\left(\frac{\rho_A}{\rho_W}\right)\frac{\mathrm{d}p_W}{\mathrm{d}x} + \left(\frac{\nu_W}{\nu_A}\right)\frac{\partial^2 u_W}{\partial \mathrm{y}^2}, \qquad (7.0.1)$$

$$\frac{\partial u_W}{\partial x} + \frac{\partial v_W}{\partial \mathrm{y}} = 0, \qquad (7.0.2)$$

The lubrication approximation enters the formulation via a perturbation approach on (7.0.1) and

(7.0.2), identical to that in Chapter 6. We rescale $\mathrm{y} = \delta Y$ and $h(x) = \delta H(x)$, for $\delta \ll 1$ and

$H, Y = O(1)$ and expect the solution in water to take the form of a power series in $\delta$:

$$u_W = \delta U_W(x, Y) + \cdots, \qquad (7.0.3)$$

$$v_W = \delta^2 V_W(x, Y) + \cdots, \qquad (7.0.4)$$

$$p_W = \frac{1}{\delta}P_W(x) + \cdots. \qquad (7.0.5)$$

The flow in air reacts to the presence of the second fluid with a non zero velocity along $h(x)$.

Before progressing further with our analysis of the droplet, the solution of the previous chapter

must be adjusted to incorporate this change.

## 7.1 Triple deck flow over a surface mounted droplet.

At the boundary between the two fluids a continuous velocity is required and (6.2.3) must equal

(7.0.3), giving the $u$ perturbation to the external fluid along the interface,

$$U_A(x, 0) = U_W(x, H(x)). \qquad (7.1.1)$$

The no slip condition along $y = 0$ (6.2.9) which was used in (6.2.17) must also be adjusted to

include (7.1.1): the constant of integration is replaced by $\widehat{U}_A(k, 0)$, which changes (6.2.18) to

$$B(k) = 3\left(\widehat{\mathcal{A}}(k) + \widehat{H}(k) - \widehat{U}_W(k, \widehat{H})\right), \qquad (7.1.2)$$

Recalculating the expressions for pressure, displacement and velocity based on (7.1.2), we find

the solution for flow over a liquid droplet as

$$\widehat{p}(k) = \frac{3|k|Ai'(0)}{|k|(ik)^{1/3} - 3Ai'(0)} \left( \widehat{H}(k) - \widehat{U}_W(k, \widehat{H}) \right),$$
(7.1.3)

$$\widehat{A}(k) = \frac{3Ai'(0)}{|k|(ik)^{1/3} - 3Ai'(0)} \left( \widehat{H}(k) - \widehat{U}_W(k, \widehat{H}) \right),$$
(7.1.4)

$$\widehat{U}_A = \frac{3|k|(ik)^{1/3}}{|k|(ik)^{1/3} - 3Ai'(0)} \left( \widehat{H}(k) - \widehat{U}_W(k, \widehat{H}) \right) \int_0^\xi Ai(s)\, ds + \widehat{U}_W(k, \widehat{H}).$$
(7.1.5)

In the following section we demonstrate that $\widehat{U}_W(k, \widehat{H})$ may be written entirely in terms of

$H(x)$, so that it is still the droplet's shape which uniquely determines the perturbation to the

flat plate problem.

This solution (7.1.3) to (7.1.5) could also be analysed numerically, an investigation we chose

not to pursue for the current problem. Instead we switch our focus to the dynamics within the

water droplet: first deriving an expression for the shape of the interface and then exploring the

families of solutions it yields in the final sections of this chapter.

## 7.2  Steady shapes of surface mounted droplets in a triple deck flow.

The fluid within the droplet has density and dynamic viscosity $\rho_W$ and $\mu_W$ respectively, both

of $O(1)$, while we assume, without loss of generality, the value of these in the outer flow to be

unity. Using (7.0.3) to (7.0.5), a balance in the governing equations for fluid within the droplet

(7.0.1), (7.0.2) yields the following:

$$\frac{\partial^2 U_W}{\partial Y^2} = \frac{1}{\mu_W} \frac{\partial P_W}{\partial x},$$
(7.2.1)

$$\frac{\partial U_W}{\partial x} + \frac{\partial V_W}{\partial Y} = 0.$$
(7.2.2)

These are as in lubrication theory. The momentum equation (7.2.1) may be integrated directly

to give an exact expression for the velocity perturbations within the droplet. A no slip condition along the solid wall, $Y = 0$, sets the second constant of integration to zero, while the other is found from the continuous shear stress across the two fluid interface $Y = H(x)$ by comparing the $y$ derivatives of (6.2.3) and (7.0.3):

$$\frac{\partial U_W}{\partial Y} = \frac{1}{\mu_W} \text{ along } Y = H(x). \tag{7.2.3}$$

This leads to a velocity described by

$$U_W = \frac{1}{2\mu_W} P_{Wx} \left( Y^2 - 2YH \right) + \frac{Y}{\mu_W}. \tag{7.2.4}$$

We may use this solution to derive two equivalent expressions for $V_W$, where equating leads to a Reynolds-lubrication type description of droplet shape. For the first, we apply the form of $U_W$ in (7.2.4) to the continuity equation (7.2.2): by differentiating with respect to $x$ and integrating with respect to $Y$ between $Y = 0$ and $Y = H(x)$. This gives us the velocity in the $y$ direction along the interface,

$$V_W = \frac{P_{Wxx}H^3}{3\mu_W} + \frac{P_{Wx}H^2 H_x}{2\mu_W}, \tag{7.2.5}$$

$$= \frac{H^{3/2}}{3\mu_W} \left[ P_{Wx}H^{3/2} \right]_x. \tag{7.2.6}$$

The second form for $V_W$ may be found from the kinematic condition for a steady two fluid interface,

$$V_A = \delta V_W = \delta H_x U_W \text{ along } Y = H(x). \tag{7.2.7}$$

We may simplify these two equivalent expressions (7.2.6) and (7.2.7), by considering the pressure jump in terms of the curvature of the surface and the surface tension across it. Adding a gravitational term, the Youngs-Laplace equation we require is

$$p - p_W = \gamma\kappa + (1 - \rho_W)gh, \tag{7.2.8}$$

for surface tension $\gamma$ and curvature

$$\kappa = \nabla \cdot \hat{n} = \frac{h_{xx}}{(1 + h_x^2)^{3/2}}. \tag{7.2.9}$$

Working under the assumption that the interface is smooth and relatively flat which is in agreement with our lubrication approximation of earlier, we may approximate the curvature $\kappa$ by the leading order term in the binomial expansion of (7.2.9): $\kappa = h_{xx} = \delta H_{xx}$. Taking (7.0.5) into account, to include both surface tension and gravity we require $\gamma, g = O(1/\delta^2)$, and so rescale as $\gamma = \gamma_W/\delta^2$, $g = g_W/\delta^2$. This leads to an $O(1/\delta)$ balance of the equation as

$$P_W = (\rho_W - 1)g_W H - \gamma_W H_{xx}. \tag{7.2.10}$$

Using (7.2.10) to eliminate the pressure in our two equivalent expressions for $V_W$, (7.2.6) and (7.2.7), we find an equation for the shape of the droplet as a function of $x$ only,

$$\gamma_W \left[ H^3 H_{xxx} \right]_x - (\rho_W - 1)g_W \left[ H^3 H_x \right]_x + \left[ \frac{3H^2}{2} \right]_x = 0. \tag{7.2.11}$$

This fourth order non linear differential equation (7.2.11) includes the effects on the droplet of surface tension (first term), gravity (second term) and drag forces exerted by the shear (third term). The family of solutions produced are fixed only by the characteristics of the droplet: its size, density and surface tension coefficient. If $H$ and $x$ are both of $O(1)$, the last two of these characteristics are of particular interest since they dictate the balance of terms and hence forces in (7.2.11). To single out the effect of these parameters, $\rho_W$ and $\gamma_W$, we restrict the range of droplets considered in our solution by prescribing the leading and trailing contact points and applying a volume constraint. This makes the boundary conditions simply

$$H(0) = 0, \qquad H(L) = 0, \tag{7.2.12}$$

$$\int_0^L H(x)\,\mathrm{d}x = V_0, \tag{7.2.13}$$

with $V_0$ and $L$ constants. Of course, no assumption limits this study to droplets, liquid films too could be analysed using this formulation. The boundary conditions are chosen to tie in with other work in the thesis.

The first condition in (7.2.12) allows us to integrate (7.2.15) once yielding a zero constant

| | $\omega \ll 1$ | Case 1: Surface tension only |
| | $\omega = O(1)$ | |
| $\gamma \gg 1$ | $\omega = O(\gamma)$ | Case 2: Surface tension and gravity balance |
| | $\omega \gg O(\gamma)$ | $H_x = 0$, No solution |
| $\gamma = O(1)$ | $\omega \ll 1$ | Case 4: Surface tension and shear forces balance |
| | $\omega = O(1)$ | Case 3: Surface tension, gravity and shear forces balance |
| | $\omega \gg 1$ | $H_x = 0$, No solution |
| $\gamma = \ll 1$ | $\omega \ll 1$ | No solution |
| | $\omega = O(1)$ | Case 5: Gravity and shear forces balance |
| | $\omega \gg 1$ | $H_x = 0$, No solution |

Table 7.1: Force balance of (7.2.15) for values of the parameters $\gamma$ and $\omega$.

of integration. For easier notation, we rewrite the difference in densities of the two fluids times

the gravity as

$$\omega = (\rho_W - 1)g_W, \tag{7.2.14}$$

a measure of the relative weight of the droplet, and drop the subscript $W$ from $\gamma$. This serves

to make our governing equation

$$\gamma H H_{xxx} - \omega H H_x + \frac{3}{2} = 0. \tag{7.2.15}$$

This expression (7.2.15) has similarities to the generalised fourth order differential equation

which describes thin film flows such as those reviewed by Myers [39] and studied by Pugh

[46]. Exploring the possible solutions to (7.2.15), subject to the boundary conditions (7.2.12)

and (7.2.13) is the focus of the remainder of this chapter. Working within the framework of this

problem it is useful to consider values of $\gamma$ and $\omega$ both larger and smaller than the $O(1)$ assump-

tion upon which (7.2.15) was derived. Although the density and surface tension coefficient are

fixed for any given fluid and setup of the liquid geometry (foam, bubble, flat surface etc), the two properties share a complex relationship, Macleod [33]. The subtleties of this relationship

are not the main concern of our study and so we assume $\gamma$ and $\omega$ to be independent. This allows

us to analyse the effect which each term (or force) has on the solution, offering insight into the

nature of the problem. Doing so is valid if we consider, for example, small values of $\gamma$ which

are still large compared to the lubrication parameter $\delta$.

Using $\gamma$ as a guide, table (7.1) details the form of equation (7.2.15) which corresponds to

the magnitude of $\omega$. The cases refer to the forces which balance in each problem, and detailed

solutions for each case follow.

It is worth pointing out at this stage that with this formulation, we may also investigate how

the forces on the droplet affect the leading and trailing contact angles. Our model is based on

a lubrication approximation, so that we require the angle the droplet makes with the wall to

be small. Returning to our triple deck coordinate system then, we expect $h'(0) = \tan\theta \approx \theta$

for $\theta << 1$, where values of $\theta < 0.4$ keep the error in neglecting subsequent terms of the

expansion below 5%. Scaling the angle at the leading contact point by $\theta = \delta\alpha$ and likewise

with $\beta$ at the trailing contact point, we find we may determine the contact angles from our

solution using

$$H'(0) = \alpha \text{ and } H'(L) = -\beta. \qquad (7.2.16)$$

where $\alpha, \beta$ are of $O(1)$.

### 7.2.1 Case 1 : Surface tension only.

When $\gamma >> 1$ and $\omega = O(1)$, shear forces and gravity have a negligible influence on droplet

shape and our equation (7.2.15) reduces to

$$H^3 H_{xxx} = 0. \qquad (7.2.17)$$

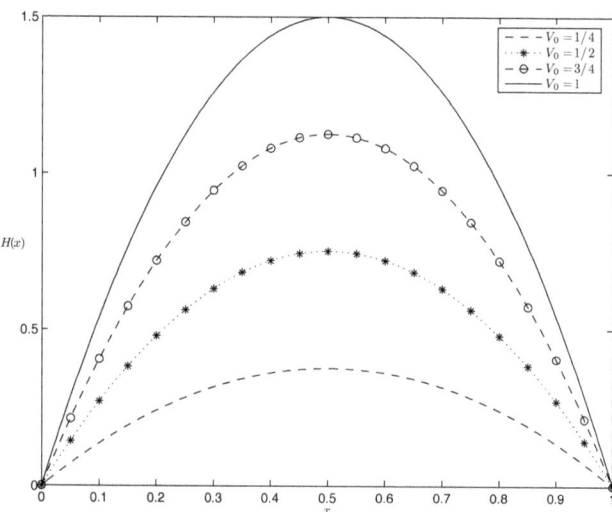

(a) Family of solutions for the surface tension only case.

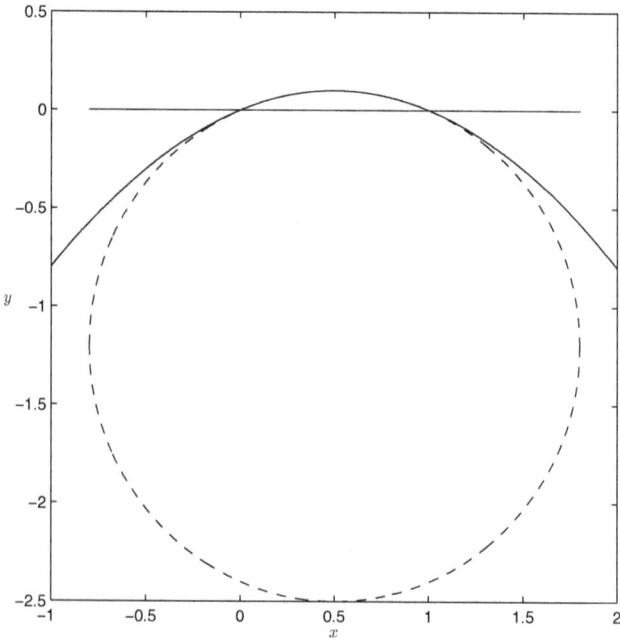

(b) Solutions to the surface tension only case form sections of a circle when considered in triple deck coordinates.

Figure 7.1: Case 1 : Surface tension only.

Application of the boundary conditions (7.2.12) and (7.2.13) lead to an exact solution:

$$H(x) = \frac{6V_0}{L^3}\left(Lx - x^2\right).$$ (7.2.18)

Since the surface tension coefficient and fluid density play no part in the trends for this case, by fixing the contact points and droplet volume we find a unique solution, symmetric about $L/2$ and with $\alpha = \beta$. This seems reasonable, since the external flow produces no net drag force upon the droplet. The contact angles may be determined by using (7.2.16), which yields a linear relationship with volume

$$\alpha = \frac{6V_0}{L^2}.$$ (7.2.19)

Example solutions may be seen in figure (7.1a). Numerically we choose to take $L = 1$ and the results show a range of droplets with volumes $V_0 = 0.25, 0.5, 0.75, 1$, corresponding to leading contact angles $\alpha = 1.5, 2, 4.5, 2$.

Taking our scaling of $y = \delta Y$ into account, we see from the figure that the droplets would appear to have a uniform curvature if considered in triple deck coordinates. This is as one would expect, since in the presence of surface tension the cohesive forces act to pull the droplet into a cylindrical shape, minimising surface area and energy state. Indeed, it can be shown that when rescaled back into the triple deck coordinates for small values of $\theta$ the solutions approximately form a section of a circle with equation

$$(x - \frac{L}{2})^2 + (y - \frac{9\delta^2 V_0^2 - L^4}{12L\delta V_0})^2 = \left(\frac{L^2}{2}\right)^2 + \left(\frac{9\delta^2 V_0^2 - L^4}{12L\delta V_0}\right)^2.$$ (7.2.20)

An example of this may be seen in figure (7.1b) where $\theta = 0.4$.

### 7.2.2   Case 2 : Surface tension and gravity.

When $\gamma \gg 1$ and $\omega = O(\gamma)$, surface tension is balanced by gravitational forces and (7.2.15) is reduced to

$$H_{xxx} = \frac{\omega}{\gamma} H_x.$$ (7.2.21)

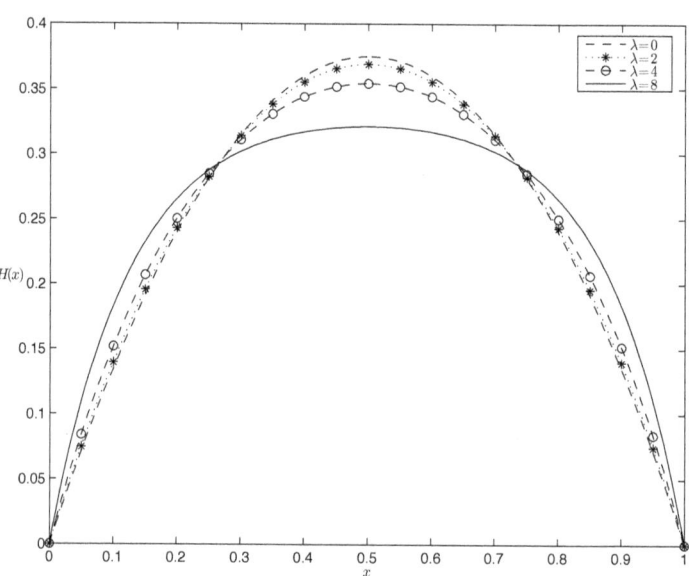

(a) Family of solutions for the case of gravity and surface tension, $V_0 = 0.25$

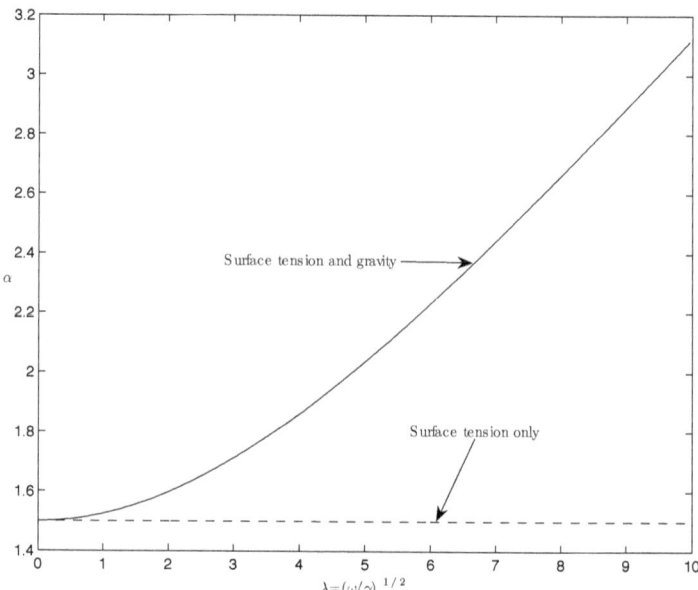

(b) The relationship between $\lambda$ and $\alpha$ becomes linear with large values.

Figure 7.2: Case 2 : Surface tension and gravity balance.

It is convenient to define a ratio between the two parameters

$$\lambda = \left(\frac{\omega}{\gamma}\right)^{1/2},$$    (7.2.22)

and consider only droplets whose density is larger than that of the surrounding fluid, so that $\lambda$

is real. Considering the form of $\omega$ in (7.2.14), $\lambda^2$ may be considered analogous to the Eötvös or

Bond number which characterises the shape of sitting droplets as described in Rosenhead [51].

As in the surface tension only case, an exact solution may be determined by application of

the boundary conditions (7.2.12) and (7.2.13):

$$H(x) = \frac{V_0\lambda\left(\tanh[\lambda L/2]\sinh[\lambda x] - \cosh[\lambda x] + 1\right)}{\lambda L - 2\tanh[\lambda L/2]},$$    (7.2.23)

The families of curves which this produces are unique for a given $\lambda$ and symmetrical about the

center of the droplet as in Case 1, which makes physical sense since there are still no shear

forces. Equation (7.2.16) again allows us to determine the linear relationship between $\alpha$ and $V_0$

for a given $L$ and $\lambda$. Specifically, this is given by

$$\alpha = \frac{V\lambda^2\tanh[\lambda L/2]}{\lambda L - 2\tanh[\lambda L/2]}.$$    (7.2.24)

Picking the first example used in figure (7.1a) as a starting point, we fix $V_0 = 0.25$, $L = 1$ and

vary $\lambda$ to explore the effect it has on $H$ and $\alpha$. Figure (7.2a) shows plots of typical solutions

for $H$ and figure (7.2b) is the relationship between $\alpha$ and $\lambda$.

For smaller values of $\lambda$ surface tension dominates and the droplet shape deviates little from the

trend seen in Case 1. (Indeed, $\lambda = 0$ is Case 1). It appears from figure (7.2b) that for values

up to $\lambda = 2$, or rather $\omega = 4\gamma$, the difference in leading contact angle compared to the surface

tension only case is slight. Beyond this point, the weight of the droplet dominates and the

droplet begins to flatten since the gravitational force acts downwards. Increasing the ratio of $\omega$

to $\gamma$ further suggests we approach a linear relationship between $\alpha$ and $\lambda$, reflected in the form

of (7.2.24).

### 7.2.3   Case 3: Surface tension, gravity and shear forces.

When $\gamma, \omega = O(1)$ we return to the full form of (7.2.15) and reintroduce the shear term:

$$\gamma H H_{xxx} - \omega H H_x + \frac{3}{2} = 0. \tag{7.2.25}$$

This non linear equation must be solved numerically, but satisfying the boundary conditions of pinned contact points (7.2.12) and fixed volume (7.2.13) cause some restrictions on the methods we may use to do so. The nature of the equation ( $H_{xxx}$ becomes infinite as $x \to 0, 1$ ) suggests that any iterative algorithm will be particulary sensitive to initial guesses for $H$. Our early computational attempts confirmed that this is indeed the case. For that reason, we propose an alternative approach, by considering the similarities to the linear form of the problem (that seen in Case 2). We suggest that the solution to (7.2.25) deviates from the linear problem by some non linear amount $A(x)$, so that $H(x)$ may be written as the sum of the two:

$$H(x) = \widetilde{H}(x) + A(x), \tag{7.2.26}$$

where $\widetilde{H}(x)$ is the known solution to the surface tension and gravity only problem (7.2.23), satisfying the boundary conditions (7.2.12), (7.2.13) and the equation (7.2.21).

Using (7.2.26) as a substitution in (7.2.25) leaves the equation to be solved as

$$\gamma A_{xxx} - \omega A_x + \frac{3}{2(\widetilde{H}(x) + A(x))} = 0. \tag{7.2.27}$$

This form contains first and third order derivatives which can be numerically sensitive. To ease this problem, we put

$$\frac{\partial A}{\partial x} = B, \tag{7.2.28}$$

so that (7.2.27) becomes

$$B_{xx} - \frac{\omega}{\gamma} B + \frac{3}{2\gamma(\widetilde{H} + A)} = 0, \tag{7.2.29}$$

which has the additional benefit of allowing us to express the boundary conditions in the form of two integrals, helping the sensitivity and convergence of any iterative algorithm. The droplet remains pinned at $x = (0, L)$ and so any deviation from the shear case, $A(x)$, must also be zero

at these points. Considering (7.2.28), this translates to the first integral condition on $B$:

$$\int_0^L B(x) = 0. \tag{7.2.30}$$

Finally, the volume constraint (7.2.13) will already be satisfied by $\widetilde{H}$, so that the integral of

$A(x)$ across the length of the droplet must also be zero,

$$\int_0^L A(x) = 0. \tag{7.2.31}$$

Equations (7.2.30) and (7.2.31) serve as our boundary conditions for the pair of equations

(7.2.28) and (7.2.29). This closed system for $A$ and $B$, with $\widetilde{H}$ known, may be solved for a

given $\omega$ and $\gamma$ with the application of an iterative algorithm similar to those seen in Part I of the

thesis.

$B$ is calculated on a first guess of $A$ using (7.2.29) and (7.2.30). The result then serves to

update the values of $A$ using (7.2.28) and (7.2.31). The process is then repeated until succes-

sive iterative values of both $A$ and $B$ converge.

As we saw in Case 2, a fixed ratio between $\gamma$ and $\omega$ implies a fixed balance between the

surface tension and gravitational forces. To understand the effect of shear forces on a given

droplet then, we choose to fix this ratio $\lambda$, and vary $\gamma$, so that $1/\gamma$ quantifies the comparative

magnitude of the shear force. Figure (7.3a) shows a plot of $A(x)$ for $\gamma = 1, 2, 4, 8$ and three

different grid spacings, $\delta x = 1/50, 1/100, 1/200$. All plots have a choice of $\lambda = 8, L = 1$ and

$V_0 = 0.25$, equivalent to applying shear to the droplet which was shown as a solid black line in

figure (7.2a).

For large values of $\gamma$ the trend for the non linear deviation $A(x)$ is approximately odd about

$L/2$. As $1/\gamma$ increases, so too does the magnitude of $A(x)$, and it appears that the point at which $A(x) = 0$ moves to the left of $L/2$. In all examples, the droplet height is decreased at the leading edge and increased at the trailing edge compared to the no shear case. This seems reasonable, since the outer fluid which causes the shear stresses is moving from left to right and matches the results of Li and Pozrikidis [31], Gupta and Basu [20] and Schliezer and Bonnecaze[52].

We notice also that the numerical algorithm holds well with grid refinement. For clarity, we include figure (7.3b) which shows solutions for $H$. In this image, the shift of the droplet to the right with increasing shear becomes clear. As $\gamma \to \infty$, $A \to 0$ we approach the solution outlined in Case 2 of $H = \widetilde{H}$, shown in the figure as a solid black line.

Despite the non linear nature of this case, it appears that we may derive an exact expression for the leading contact angle by considering the integral of (7.2.25). Since (7.2.25) may be rewritten

$$\gamma \left[ H H_{xx} \right]_x - \frac{\gamma}{2} \left[ H_x^2 \right]_x - \frac{\omega}{2} \left[ H^2 \right]_x + \frac{3}{2} = 0, \tag{7.2.32}$$

the value of the integral of (7.2.32) at $x = 0$,

$$-\frac{H_x^2(0)}{2} = C, \tag{7.2.33}$$

combined with that at $x = L$,

$$-\frac{H_x^2(L)}{2} + \frac{3L}{2} = C \tag{7.2.34}$$

for some constant $C$, yields an equality.

$$H_x(L)^2 - H_x(0)^2 = \frac{3L}{\gamma}. \tag{7.2.35}$$

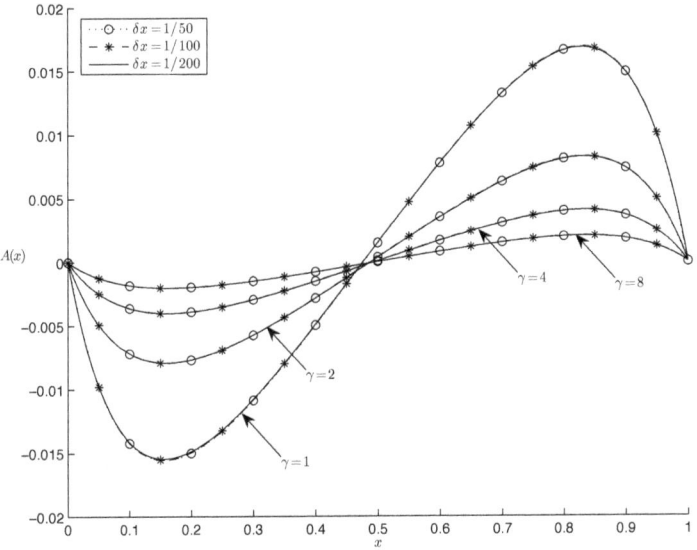

(a) The deviation of droplet shape from the shear case

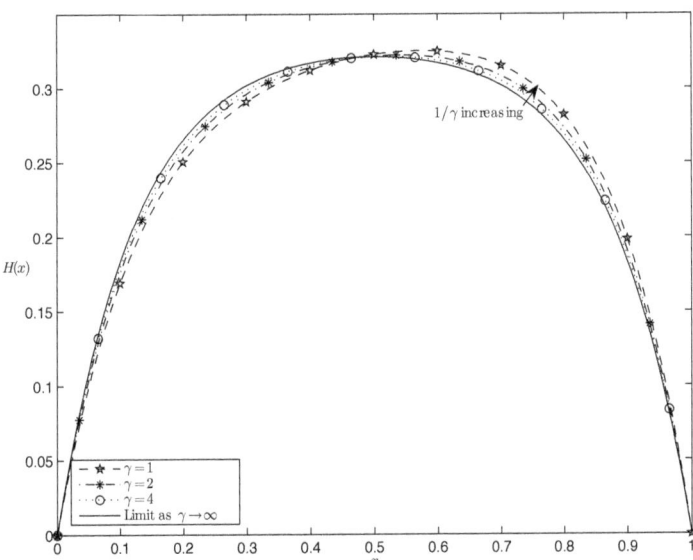

(b) Droplet shapes for the full non linear equation

Figure 7.3: Case 4 : Surface tension, gravity and shear balance.

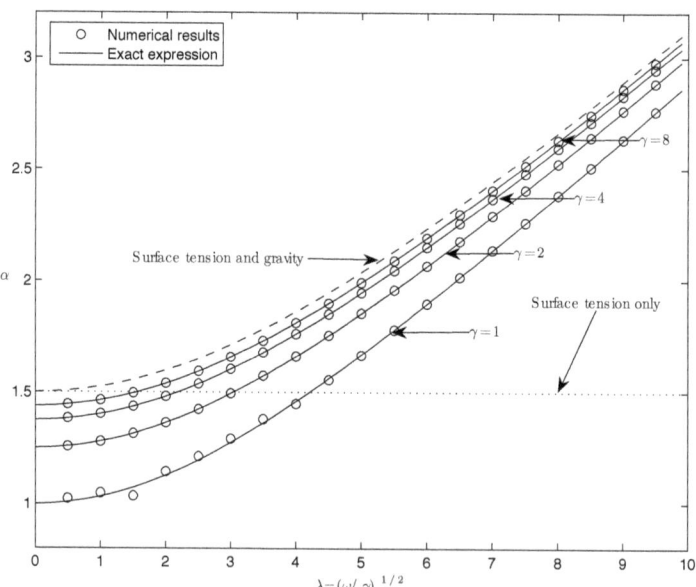

Figure 7.4: The effect of force balances on leading contact angle for the full problem: surface tension, gravity and shear forces.

Rewriting (7.2.35) in terms of $H = \widetilde{H} + A$ and, since $\widetilde{H}$ is the solution to the problem with no shear, $\widetilde{H}_x(0) = -\widetilde{H}_x(L)$, we find

$$\left(A_x(L)^2 - A_x(0)^2\right) - 2\widetilde{H}_x(0)\left(A_x(L) + A_x(0)\right) = \frac{3L}{\gamma}. \tag{7.2.36}$$

It appears from figure (7.3a) that the gradients of $A(x)$ as $x \to 0$ and $x \to L$ are similar. If they were equal, $A_x(0) = A_x(L)$, (7.2.36) would suggest that

$$A_x(0) = A_x(L) = -\frac{3L}{4\gamma\widetilde{H}_x(0)}. \tag{7.2.37}$$

An exact solution for $\widetilde{H}_x(0)$ is known and given in (7.2.24) (which we shall refer to as $\alpha_0$ from now on) so that if (7.2.37) is true, we may write down an expression for the leading and trailing contact angles

$$\alpha = H_x(0) = \widetilde{H}_x(0) + A_x(0) = \alpha_0 - \frac{3L}{4\gamma\alpha_0}, \tag{7.2.38}$$

$$\beta = H_x(0) = \widetilde{H}_x(L) + A_x(L) = -\alpha_0 - \frac{3L}{4\gamma\alpha_0}. \tag{7.2.39}$$

We may compare these expressions (7.2.39) to our numerical solution. Results may be seen in figure (7.4), an extension of our earlier plot, figure (7.2b) of leading contact angle $\alpha$ against $\lambda$. Each solid black curve shows the leading contact angle for a fixed $\gamma$ and a range of $\lambda$, calculated using (7.2.38). We also include the computational results in the figure as black circles, found from (7.2.28) - (7.2.31), where the angle is calculated from the first three grid points.

The figure (7.4) demonstrates that (7.2.39) matches well to the numerical results, aside from values of $\lambda < 4$ for $\gamma = 1$ where it was found relaxation was required to arrive at a solution. This corresponds to small values of $\omega$ on which we suspect there may be some restriction hidden within the non linearity of the system. In Case 5 which follows we find a similar restriction and demonstrate that the magnitude of $\omega$ acts to resist a shear force. Since these anomalies refer to droplets which have a low gravitational force and are subjected to moderately high shears, it seems reasonable to propose we see the beginning, or limit, of the droplet sliding or rolling.

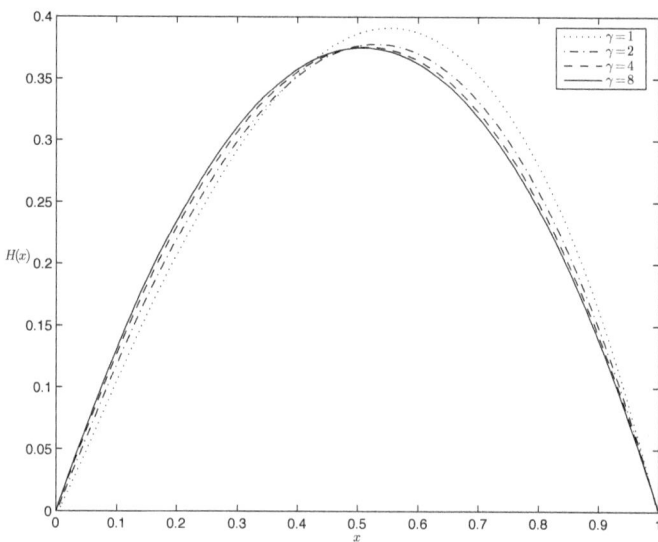

Figure 7.5: Case 4 : Family of solutions for the case of surface tension and shear balance.

We therefore have some confidence in suggesting that $A_x(0)$ does indeed equal $A_x(L)$ and conclude that while the leading and trailing contact angles differ, their deviation from the no shear case does not. The droplet is not symmetrical in the main however; the shape as $x \to 0$ and $x \to L$ is tilted by the same degree for a given shear.

### 7.2.4  Case 4 : Surface tension and shear forces.

When $\omega << 1$ and $\gamma = O(1)$ shear forces are balanced by surface tension and (7.2.15) becomes

$$\gamma H H_{xxx} = -\frac{3}{2}. \tag{7.2.40}$$

We may think of this as a special example of the non linear equation seen in Case 3, and apply the same numerical algorithm with $\omega = 0$ to solve it. Results for $\gamma = 1, 2, 4, 8$ are shown in figure (7.5), where values of $\delta x = 1/200$ , $L = 1$, $V_0 = 0.25$ are taken throughout the calculations. We notice a similar trend to that seen in figure (7.3b) with the droplet shifting to the right as $1/\gamma$ increases. As $\gamma \to \infty$ the limit of this droplet is the surface tension only

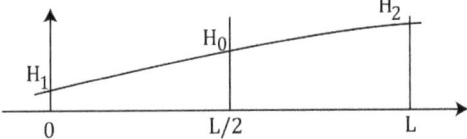

Figure 7.6: A diagram of the droplet core $) < x < L$ for the case of a gravity and shear force balance. End layers arise near $x = 0, L$.

problem seen in Case 1.

It was found to be difficult to obtain results for $\gamma < 1$ which supports our earlier suggestion that gravity must be present for the droplet to resist large drag forces exerted by the shear and remain attached to the wall.

### 7.2.5 Case 5: Gravity and shear forces.

When $\gamma << 1$ and $\omega = O(1)$ the equation for droplet shape (7.2.15) reduces to

$$\omega H H_x = \frac{3}{2}. \tag{7.2.41}$$

This equation may be integrated directly, but cannot satisfy the boundary conditions at $x = 0$ and $x = L$ without the influence of surface tension near these points. As we have seen in Case 2, gravity acts to flatten the droplet. With $\gamma = 0$ everywhere, there would be no force acting to pull the droplet into a cylindrical shape and fix the contact points. To solve for this case analytically, we must consider three parts to the solution: the core problem governed by (7.2.41) and two end layers, a description of which is given below. A diagram of this setup is given in figure (7.6). $H_1$ and $H_2$ are the droplet heights at $x = 0$ and $x = L$ which the core solution provides. The end layers must match to these values as $x$ increases from zero and decreases from $L$ respectively. If the droplet has height $H_0$ at $x = L/2$ we find the expression for the droplet's shape in the core:

$$H = \left( \frac{3}{2\omega} (2x - L) + H_0^2 \right)^{1/2}, \tag{7.2.42}$$

where the value of $H_0$ is fixed by the volume constraint (7.2.13) and must satisfy

$$\left( H_0^2 + \frac{3}{2\omega}L \right)^{3/2} - \left( H_0^2 - \frac{3}{2\omega}L \right)^{3/2} = \frac{9V_0}{2\omega}, \tag{7.2.43}$$

clearly suggesting a restriction on the values of $\omega$

$$\omega \geq \frac{3L}{2H_0^2}. \tag{7.2.44}$$

Plots of (7.2.43) may be seen in figure (7.7a). The curves correspond to lines of constant $H_0$ and demonstrate the relationship between $\omega$ ($x$ axis) and $V_0$ ($y$ axis) with $L$ taken to be unity throughout. To tie in to other cases, we require a choice of $H_0$ and $\omega$ which yields $V_0 = 0.25$, shown on our figure as a dashed line. We may pick any $H_0$, $\omega$ pair which intersects with this line and so choose $A$, $B$ and $C$ as labelled in the figure . Of course, there will be an infinite number of pairs which will yield such a result for $V_0$, but all will lie within the range $0.25 < H_0 < 0.27$ and hence $\omega > 20$. Perhaps this is a surprisingly high minimum value for $\omega$, suggesting that only fluids which are very dense compared to the surrounding flow can form this type of droplet and resist a shear force without surface tension in the core.

For these choices $A$, $B$ and $C$ we also include figure (7.7b), showing plots of the corresponding solutions to (7.2.41) which demonstrate the effect which $\omega$ has on the core of the droplet. As suggested by the equations, the size of $\omega$ is inversely related to $H_0$: a droplet with a smaller gravitational force ratio will result in a higher curvature as $x \to 0$ and a greater difference between $H_1$ and $H_2$.

For the end layer at the leading edge of the droplet we return to the full form of (7.2.15) and assume expressions for $x$ and $H$ as

$$x = \gamma^{1/2}\bar{x}, \tag{7.2.45}$$

$$H = \bar{H} + \cdots . \tag{7.2.46}$$

This serves to enforce a surface tension-gravity balance at the leading contact point. Applying

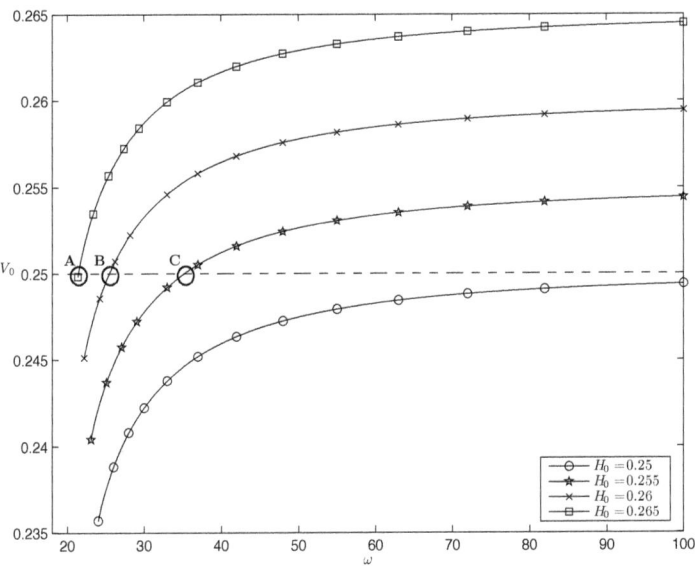

(a) Restrictions on the choice of $\omega$ for the case of a shear and gravity balance

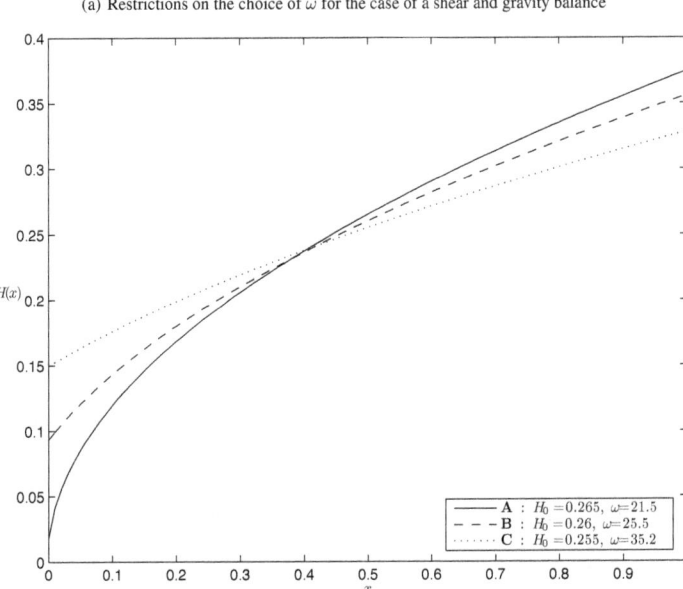

(b) Example plots for the core solution in the case of a shear and gravity balance.

Figure 7.7: Case 5: Gravity and shear balance

(7.2.12) and matching to $H_1$ as $\bar{x} \to \infty$ we find the solution to be

$$\bar{H} = H_1 \left( 1 - \exp\left[ -\omega^{1/2}\bar{x} \right] \right). \tag{7.2.47}$$

A similar procedure may be applied at $x = L$, where the expression for $x$ takes the form

$$x = L - \gamma^{1/2}\hat{x} + \cdots, \tag{7.2.48}$$

and the solution for $H$ is given by

$$\hat{H} = H_2 \left( 1 - \exp\left[ -\omega^{1/2}\hat{x} \right] \right). \tag{7.2.49}$$

Since surface tension is required to satisfy the boundary conditions we can return to our computational algorithm used in Case 4 and compare numerical results for small $\gamma$, and the values of $\omega$ in $A, B$ and $C$ already discussed, with this analytical solution. Figure (7.7b) may then be thought of as the limit of these solutions as $\gamma \to 0$.

Plots of the results for $\gamma = 0.3$ and $\gamma = 0.1$ are shown in figures (7.8a) and (7.8b) respectively. The line styles in the figures correspond to those in figure (7.7b) ($A$ represented by a solid line and so on). We see that the trend is the same as before: the heavier droplets have the ability to resist high shear forces and appear flatter and more symmetrical. The effect of $\gamma$ (and hence surface tension) is to increase the curvature throughout, demonstrated by the differences between figure (7.8a) and figure (7.8b).

By considering the problem in this manner ($\gamma$ small rather than $\gamma \to 0$) the restrictions on $\omega$ found earlier perhaps become clearer. To withstand high shears and remain steady a droplet must have a medium to large gravitational force or there is nothing to prevent the droplet from sliding.

This also sheds light on the findings of Case 3, where results for small gravitational forces

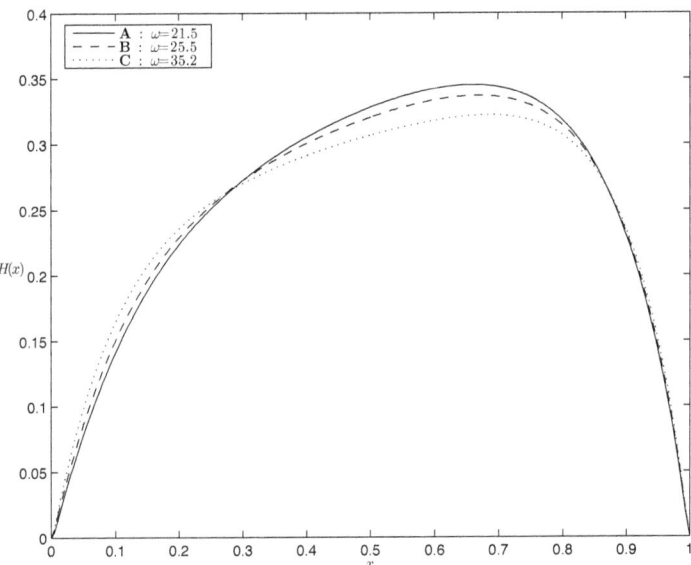

(a) Example droplet shapes for when shear and gravity dominate. Here $\gamma = 0.3$.

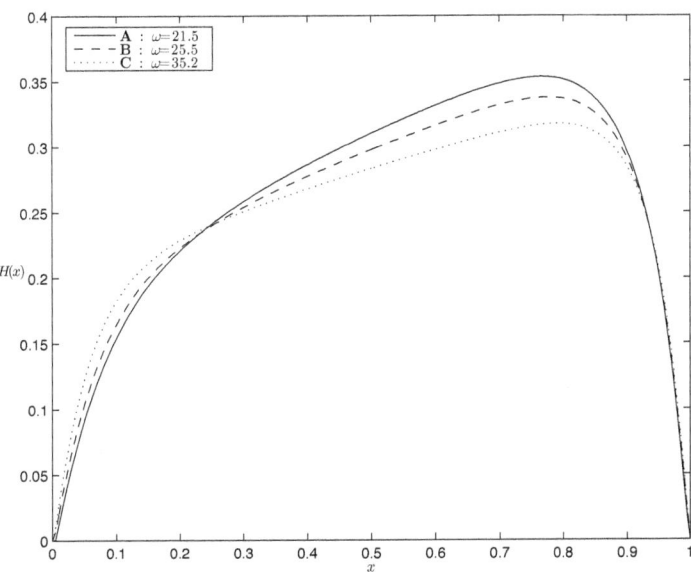

(b) Example droplet shapes for when shear and gravity dominate. Here $\gamma = 0.1$.

Figure 7.8: Case 5: Dominant Gravity and shear

$\omega$ in the presence of shear were difficult to find. If the numerical method converged, it required relaxation to do so.

This approach then, while offering insight into the effects of surface tension, gravity and shear forces breaks down with the onset of rolling, sliding or spreading and if a steady state does not exist.

To better understand the physical behaviour of a droplet within a boundary layer, and particularly it's movement over time, we must return to the small density ratios approach and tackle our final task: the two-way interacting problem of a liquid droplet deforming within the boundary layer of an external air flow.

## Chapter 8

# Droplet deformation within a boundary layer

Although several complementary problems have also been included, the main focus of this thesis has been in modelling the time dependent distortion of the interface between a surface mounted droplet and an external flow field. In this chapter we return to that goal with our final example problem: the nonlinear interaction of air and water contained within a triple deck structure, a schematic diagram of which may be seen in figure (6.1).

To examine the deformation of the water droplet, we work under the assumption of a small density ratio, details of which may be seen in Chapter 1. It is the small density ratio which allows us to treat the droplet as quasi solid and apply a no-slip condition in the air along the boundary between the two fluids. This is the basis of the method for modelling droplet deformation, also outlined in Chapter 1, which remains unchanged for the current problem.

With this approach, the two fluids, air (A) and water (W), are governed by the non-dimensional equations (1.2.13)- (1.2.16), which we repeat here for clarity:

$$\frac{\partial \mathbf{u}_A}{\partial t} + (\mathbf{u}_A \cdot \nabla)\, \mathbf{u}_A = -\nabla p_A + \left(\frac{\nu_A}{\nu_W}\right) \mathrm{Re}_W^{-1} \nabla^2 \mathbf{u}_A, \qquad (8.0.1)$$

$$\nabla \cdot \mathbf{u}_A = 0, \qquad (8.0.2)$$

$$\frac{\partial \mathbf{u}_W}{\partial t} = -\nabla p_W + \mathrm{Re}_W^{-1} \nabla^2 \mathbf{u}_W, \qquad (8.0.3)$$

$$\nabla \cdot \mathbf{u}_W = 0. \tag{8.0.4}$$

We notice the similarities between the governing equations of air in the small ratios problem (8.0.1), (8.0.2), and those of air, nondimensionalised using the properties of its own flow in Chapter 6 , equations (6.0.2), (6.0.3), where a solution to the air flow over a solid obstacle was derived. Indeed, by rewriting

$$\widehat{\mathrm{Re}} = \left( \frac{\nu_W}{\nu_A} \right) \mathrm{Re}_W, \tag{8.0.5}$$

the two sets are identical except that in the present case $Re_A$ is replaced below by $\widehat{\mathrm{Re}}$. Assuming that $\widehat{\mathrm{Re}}$ is large, all steady state findings and triple deck scalings of Chapter 6 will apply here also.

As with the example in Part I of the thesis, the deformation of a droplet within a boundary layer occurs across two temporal scales. In the early stage, while $t = O(1)$ the unsteady flow within the water droplet is driven by the unsteady flow in air. As $t$ increases, both flows become steady and the droplet begins to grow linearly with time. While $t < O(1/\epsilon)$ the droplet retains its initial shape to leading order and the interaction is considered one way.

When $t$ becomes of $O(1/\epsilon)$, the time dependent term begins to dominate the shape of the interface and the droplet becomes severely distorted, causing the kinematic condition to become nonlinear. This corresponds to the later temporal stage, where a rescaling of $t = T/\epsilon$ for $T = O(1)$ is required. During this phase, the shape change of the droplet acts to modify the dynamics within the air and the interaction is considered two-way.

The early stage was studied in the first part of the thesis in the case of a semicircular droplet, where the unsteady flows in both air and water were found separately. For the current problem, we do not provide a solution for the unsteady flow in air, focusing instead on analytical solutions for the flow within the droplet, subject to some known interfacial stress conditions. This early stage solution is presented in Section 8.1.

As $t$ becomes large, the flow in both fluids becomes steady. To determine the steady flow in air, the solution of Chapter 6 may be adapted include a small ratios approach rather than a lubrication approximation. This adapted algorithm is the task of Section 8.2, whose results are valid for large $t$ in the early stage, and in the later stage itself.

By considering the solution in the air as quasi-steady solution over the latest interface shape at every time step we may tackle the two-way nonlinear interacting problem of the later temporal stage. In Section 8.3 we derive a Reynolds Lubrication type equation which describes the shape of the interfacein terms of the properties of the air flow. Finally, in Section 8.4 some numerical simulations are presented for the flow over various initial droplet shapes.

## 8.1 The early temporal stage

The flow in water is governed by the Stokes equations at the beginning of the chapter in equations (8.0.3) and (8.0.4). Since the droplet is contained within the viscous sublayer of the triple deck structure, the inner deck scalings $\widehat{Re}^{-3/8}$ in the horizontal coordinate and $\widehat{Re}^{-5/8}$ in the vertical will apply along with

$$u_W = \widehat{Re}^{-1/8} \tilde{u}_W(X, Y) + \cdots, \tag{8.1.1}$$

$$v_W = \widehat{Re}^{-3/8} \tilde{v}_W(X, Y) + \cdots, \tag{8.1.2}$$

$$p_W = \widehat{Re}^{-1/4} \tilde{p}_W(X) + \cdots. \tag{8.1.3}$$

To include the effects of time, $t$ is scaled as $t = \widehat{Re}^{-2/8} \tilde{t}$. The tildes are dropped for ease of notation and since Prandtl's shift is not used for the droplet, we name the scaled transverse coordinate here as y, related to that in air by $y = y + f(x)$ (where $f(x)$ is the obstacle's boundary).

By applying these scalings to the momentum equation (8.0.3), the govening equations for the flow within the water droplet reduce to:

$$\frac{\partial u}{\partial t} = -\frac{\partial p}{\partial x} + \left(\frac{\nu_A}{\nu_W}\right)\frac{\partial^2 u}{\partial y^2}, \tag{8.1.4}$$

$$\frac{\partial u}{\partial x} + \frac{\partial v}{\partial y} = 0. \tag{8.1.5}$$

As briefly discussed, we will assume for the purposes of the analysis within this section, that a time dependent solution in air over an obstacle is known and concern ourselves solely with the flow in water.

This problem corresponds to the work of Chapter 2. In this example however, unlike that of the semicircle in Chapter 2, the leading order stress conditions are equivalent to the leading order pressure and vorticity terms. Our results appear to suggest that here, any linear growth in velocities will saturate, as postulated in Chapter 4. In the calculations which follow we see the velocities becoming steady with time.

In the early stage, the interface shape takes the form

$$y = f_0(x) + \epsilon f_1(x, t), \tag{8.1.6}$$

and we take the stress conditions along the boundary of the two fluids as

$$\frac{\partial u}{\partial y} = \sigma_0, \quad p = p_0 \quad \text{on } y = f, \tag{8.1.7}$$

to leading order, for some known values $\sigma_0$, $p_0$. A no slip condition, $u = 0$ on y= 0 completes the problem description.

The solution is relatively straightforward; the momentum equation within the droplet may be differentiated with respect to y to yield the one-dimensional heat equation for the leading order vorticity terms:

$$\left(\frac{\nu_A}{\nu_W}\right)\frac{\partial^2 \zeta}{\partial y^2} = \frac{\partial \zeta}{\partial t}. \tag{8.1.8}$$

Assume that the vorticity and its boundary conditions decay exponentially to a steady state, so that they take the form:

$$\zeta(x, y, t) = z_0(x, y) + z_1(x, y, t) \tag{8.1.9}$$

$$\zeta(x, f, t) = g(x) + h(x, t), \tag{8.1.10}$$

$$\zeta(x, 0, t) = q(x) + r(x, t), \tag{8.1.11}$$

where $g, h, q, r$ are known from (8.1.7), with (8.1.11) replacing the condition on pressure, and $z_1, h, r$ decay exponentially with time.

The steady state problem ((8.1.8 - (8.1.11) with $z_1, h, r = 0$) may be solved directly to give

$$z_0 = q(x)\,(y - f(x)) + g(x). \tag{8.1.12}$$

By applying (8.1.12) to the unsteady problem, (8.1.8 - (8.1.11), it remains to solve for $z_1$:

$$\left(\frac{\nu_A}{\nu_W}\right)\frac{\partial^2 z_1}{\partial y^2} = \frac{\partial z_1}{\partial t}, \tag{8.1.13}$$

$$z_1(x, f, t) = h(x, t), \tag{8.1.14}$$

$$\frac{\partial z_1(x, 0, t)}{\partial y} = r(x, t). \tag{8.1.15}$$

If $h(x, t)$ and $r(x, t)$ may be written as in terms of a sine series,

$$h(x, t) = \sum_n h_n \exp[-\left(\frac{\nu_A n^2 \pi^2 t}{\nu_W}\right)]\sin n\pi x, \tag{8.1.16}$$

$$r(x, t) = \sum_n r_n \exp[-\left(\frac{\nu_A n^2 \pi^2 t}{\nu_W}\right)]\sin n\pi x. \tag{8.1.17}$$

then $z_1$ will have solution

$$z_1 = \sum_n \exp[-\left(\frac{\nu_A n^2 \pi^2 t}{\nu_W}\right)]\sin n\pi x\,(a_n \sin n\pi y + b_n \cos n\pi y). \tag{8.1.18}$$

Finally, applying (8.1.16), (8.1.17) yields the values of $a_n$ and $b_n$.

$$a_n = \frac{h_n}{\sin n\pi f} + \frac{r_n}{n\pi}, \tag{8.1.19}$$

$$b_n = -\frac{r_n}{n\pi}\tan n\pi f. \tag{8.1.20}$$

The expressions (8.1.12), (8.1.18), (8.1.19) (8.1.20) together form an exact solution for the

vorticity within the droplet,

$$\zeta = q(x)\,(y - f(x)) + g(x) + \tag{8.1.21}$$

$$+ \sum_n e^{-\left(\frac{\nu_A n^2 \pi^2 t}{\nu_W}\right)} \sin n\pi x \left\{ \left(\frac{h_n}{\sin n\pi f} + \frac{r_n}{n\pi}\right) \sin n\pi y - \frac{r_n}{\pi} \tan n\pi f \cos n\pi y \right\}.$$

Since $\zeta = \dfrac{\partial u}{\partial y}$, a solution for the velocities may also be found. All terms become steady and

this example sees no unbounded linear growth of velocity with time.

Coupling this expression for vorticity (8.1.21) with a time dependent solution in air over a

quasi-solid obstacle would provide a complete description of the early stage of droplet de-

formation, as we had in Chapter 4. We do not attempt this for the current problem however,

choosing instead to focus our attention on the later stage and nonlinear two-way interaction,

where our first task is to adapt the flow in air algorithm of Chapter 6

## 8.2 The later temporal stage: flow in air

Although in the later stage, the triple deck scaled shape $f$ will have a relatively slow depen-

dence on time, the present setting of droplet deformation allows us to treat the flow in air as

quasi steady. We would like to extend the solution of Chapter 6 however (the setup of which

is valid here with $\mathrm{Re}_A$ replaced by $\widehat{\mathrm{Re}}$) before applying it as the first step in modelling droplet

deformation, by outlining a nonlinear solution and removing the shallow droplet assumption.

We do so by replacing the perturbation approach, which was used to derive (6.2.3) - (6.2.6),

and instead consider the following expansion:

$$u_A = y + U_A, \tag{8.2.1}$$

$$v_A = V_A, \tag{8.2.2}$$

$$p_A = P_A, \tag{8.2.3}$$

where $U_A, V_A, P_A$ are all $O(1)$. Application of the new expansion (8.2.1) - (8.2.3) to the relevant form of the steady inner deck equation (6.1.11) leaves

$$y\frac{\partial U_A}{\partial x} + V_A + \frac{\mathrm{d}P_A}{\mathrm{d}x} - \frac{\partial^2 U_A}{\partial y^2} = -U_A\frac{\partial U_A}{\partial x} - V_A\frac{\partial U_A}{\partial y}, \tag{8.2.4}$$

where $x, y$ are as defined in the inner deck equations, after Prandtl's transposition: see Section 6.2.

By replacing the droplet with an obstacle, the continuity equation and boundary conditions (6.2.8) - (6.2.12) also apply here, except that of course $U_A, V_A, P_A$ are now as defined in (8.2.1) - (8.2.3).

The non linear terms in (8.2.4) have been grouped on the right hand side of the equality. By setting them to zero we return to a linear solution identical in form to that seen in (6.2.21) - (6.2.23), which we rename here with a superscript 0:

$$\widehat{P}_A^0(k) = \frac{3|k|Ai'(0)}{|k|(ik)^{1/3} - 3Ai'(0)}\widehat{f}(k), \tag{8.2.5}$$

$$\widehat{A}^0(k) = \frac{3Ai'(0)}{|k|(ik)^{1/3} - 3Ai'(0)}\widehat{f}(k), \tag{8.2.6}$$

$$\widehat{U}_A^0(k, \xi) = \frac{3|k|(ik)^{1/3}}{|k|(ik)^{1/3} - 3Ai'(0)}\widehat{f}(k)\int_0^\xi Ai(s)\,\mathrm{d}s. \tag{8.2.7}$$

For a given obstacle shape $f(x)$, $\widehat{f}(k)$ may be found using the forward Fourier transform given in (6.2.13). The inverse transform of (8.2.5), (8.2.7) is then easily found numerically using the discrete sum (6.2.31), yielding real solutions which we call $(U_A^0, P_A^0)$.

We feed this linear solution back into the momentum equation (8.2.4) and iterate into the non linearity. For ease of notation, we rename the (now known) nonlinear terms as

$$g^0(x, y) = -U_A^0 \frac{\partial U_A^0}{\partial x} - V_A^0 \frac{\partial U_A^0}{\partial y} \tag{8.2.8}$$

where $V_A^0$ is found from the continuity equation.

In general, we may find the next solution for the velocity $U_A^{n+1}$ by knowing the latest $g^n$,

$$y \frac{\partial U_A^{n+1}}{\partial x} + V_A^{n+1} + \frac{\mathrm{d} P_A^{n+1}}{\mathrm{d} x} - \frac{\partial^2 U_A^{n+1}}{\partial y^2} = g^n. \tag{8.2.9}$$

The solution follows the method in Chapter 6, differentiating (8.2.9) with respect to $y$, taking

the forward Fourier transform and putting $\xi = (ik)^{1/3} y$, so that in this case, it remains to solve

the inhomogenous Airy equation, with a known forcing:

$$\xi \frac{\partial \widehat{U}_A^{n+1}}{\partial \xi} - \frac{\partial^2 \widehat{U}_A^{n+1}}{\partial \xi^2} = (ik)^{-2/3} \frac{\partial \widehat{g}^n}{\partial \xi}. \tag{8.2.10}$$

This partial differential equation (8.2.10) suggests a form for $\widehat{U}_A^{n+1}$ which is a combination of

the linear solution and an inhomogeneous contribution, and we seek a solution which follows:

$$\frac{\partial \widehat{U}_A^{n+1}}{\partial \xi} = \frac{\partial \widehat{U}_A^0}{\partial \xi} + Ai(\xi) \alpha^n(\xi). \tag{8.2.11}$$

Applying this to (8.2.10) suggests it remains to solve

$$\frac{\partial}{\partial \xi} \left( \frac{\partial \alpha^n}{\partial \xi} Ai(\xi)^2 \right) = (ik)^{-2/3} Ai(\xi) \frac{\partial \widehat{g}^n}{\partial \xi}, \tag{8.2.12}$$

$$\alpha^n(\xi) = 0 \quad \text{at} \quad \xi = 0, \tag{8.2.13}$$

$$Ai(\xi) \alpha^n(\xi) \to 0 \quad \text{as} \quad \xi \to \infty, \tag{8.2.14}$$

where (8.2.12) can be integrated directly:

$$\alpha^n(\xi) = \int_0^\xi Ai(t)^{-2} \int_\infty^t (ik)^{-2/3} Ai(s) \frac{\partial \widehat{g}^n}{\partial s} \, \mathrm{d}s \, \mathrm{d}t. \tag{8.2.15}$$

The limits on the second integral are set to ensure that (8.2.14) is met. When $g^n$ is known then,

(8.2.15) allows us to numerically calculate $\alpha^n$, which in turn provides the solution for $\widehat{U}_A^{n+1}$:

$$\widehat{U}_A^{n+1} = \widehat{U}_A^0 + \int_0^s Ai(s) \alpha^n(s) \, \mathrm{d}s. \tag{8.2.16}$$

Finally, the derivative of (8.2.11), matched to the Fourier transform of the momentum equation

(8.2.4) along $y = 0$ gives the solution for the pressure:

$$\widehat{P}_A^{n+1} = (ik)^{-1/3} \left( \left.\frac{\partial^2 \widehat{U}_A^0}{\partial \xi^2}\right|_{\xi=0} + Ai(0) \left.\frac{\partial \alpha^n}{\partial \xi}\right|_{\xi=0} + Ai'(0)\alpha^n(0) \right). \qquad (8.2.17)$$

So, from $g^0$, $U_A^1$ and $P_A^1$ may be found via $\alpha^1$, subsequently yielding $g^1$ and so on. This

procedure may be repeated until successive calculated values of $U_A$ and $P_A$ converge.

To demonstrate the method, we return to the Witch of Agnesi and parabolic shaped obstacles

which were used as examples in Chapter 6. Figure (8.1) has the nonlinear pressure and wall

shear results for the parabolic hump, calculated on the same $x$ grid, $\delta x = 10^{-3}$, $\delta y = \dfrac{\delta x^{1/2}}{4}$,

and period $L = 6\pi$ as the linear solution in figure (6.5). Again, the number of Fourier modes

considered, $N_k$, is varied from 16 to 512. We notice improvement as $N_k$ increases, although

in this nonlinear example, the solution has not yet settled completely by $N_k = 512$, showing

some sinusoidal behaviour upstream of the droplet in the wall shear image.

In figure (8.2) we show a comparison between the linear solution for pressure and wall shear,

found in Chapter 6, and the nonlinear solution outlined here for the case of the Witch of Agnesi

obstacle. We see that in general linear theory holds well for this example, deviating from the

nonlinear solution only at the extremal values of pressure and shear.

## 8.3   Later temporal stage: flow in water

As the later stage comes into operation, $t$ is rescaled as $t = T/epsilon$ with $T = O(1)$.

Application of this rescaling to (8.1.4) leaves the governing equation in water as

$$\frac{\partial^2 u_W}{\partial y^2} = \left(\frac{\nu_W}{\nu_A}\right) \frac{\partial p_W}{\partial x}. \qquad (8.3.1)$$

As in Section 8.1, (8.3.1) is solved subject to the full stress conditions, which, when non-

dimensionalised on the characteristics of the water and scaled according to both small ratios

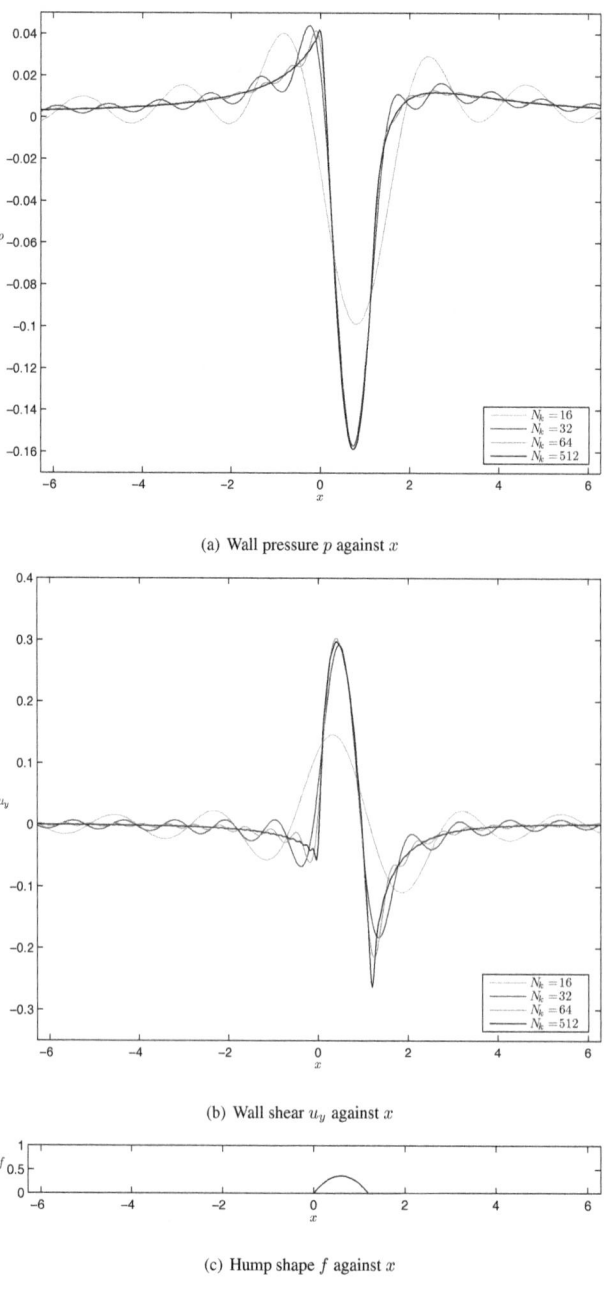

(a) Wall pressure $p$ against $x$

(b) Wall shear $u_y$ against $x$

(c) Hump shape $f$ against $x$

Figure 8.1: Nonlinear results for the parabolic hump

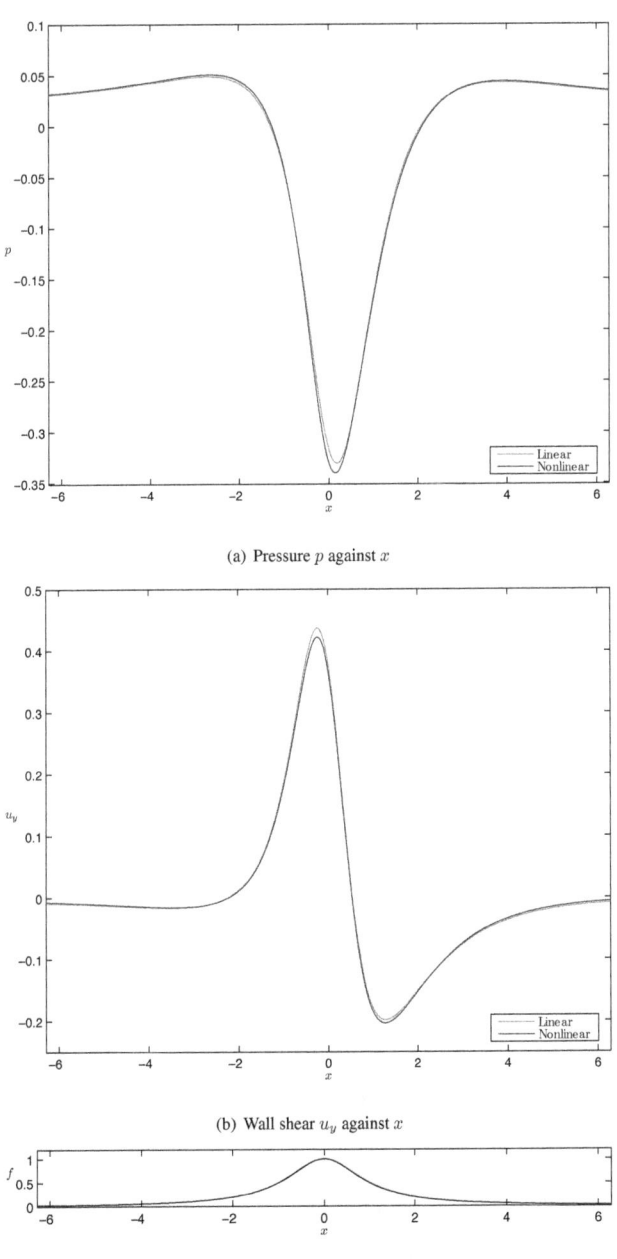

(a) Pressure $p$ against $x$

(b) Wall shear $u_y$ against $x$

(c) Hump shape $f$ against $x$

Figure 8.2: Nonlinear results for the witch of Agnesi

assumption and inner deck parameters simplify to a requirement of continuous pressure perpendicular to the interface,

$$P_A = p_W \quad \text{along} \quad y = f(x), \ y = 0, \tag{8.3.2}$$

while tangential to the interface,

$$\frac{\partial u_W}{\partial y} = \left(\frac{\nu_A}{\nu_W}\right)\left(1 + \frac{\partial U_A}{\partial y}\right) \quad \text{along} \quad y = f(x), \ y = 0, \tag{8.3.3}$$

must be satisfied. These correspond to $p_0$ and $\sigma_0$ of Section 8.1 respectively. Both (8.3.2) and (8.3.3) are known from the flow solution in air of the previous section.

Renaming the shear term in (8.3.3):

$$\sigma = \left(1 + \frac{\partial U_A}{\partial y}\right), \tag{8.3.4}$$

applying a no slip condition along $y = 0$ and the interface conditions (8.3.2) and (8.3.3), the momentum equation (8.3.1) may be solved exactly to yield a solution for the velocity within the droplet,

$$u_W = \left(\frac{\nu_W}{\nu_A}\right)\frac{1}{2}\frac{\partial p_W}{\partial x}\left(y^2 - 2yf(x)\right) + \left(\frac{\nu_A}{\nu_W}\right)\sigma y. \tag{8.3.5}$$

Using the same approach as that which derived (7.2.11) in Chapter 7, we may use (8.3.5) to find two equivalent expressions for $v_W$ along the interface, the first from the continuity equation:

$$v_W = \frac{\nu_W}{\nu_A}\left(\frac{f^3}{3}\frac{\partial^2 p}{\partial x^2} + \frac{f^2}{2}\frac{\partial p}{\partial x}\frac{\partial f}{\partial x}\right) - \left(\frac{\nu_A}{\nu_W}\right)\frac{f^2}{2}\frac{\partial \sigma}{\partial x} \tag{8.3.6}$$

and the second from the kinematic condition:

$$v_W = u\frac{\partial f}{\partial x} + \frac{\partial f}{\partial t} = \left(\frac{\nu_W}{\nu_A}\right)\frac{f^2}{2}\frac{\partial p_W}{\partial x}\frac{\partial f}{\partial x} + \left(\frac{\nu_A}{\nu_W}\right)\sigma f\frac{\partial f}{\partial x} + \frac{\partial f}{\partial t}. \tag{8.3.7}$$

Equating the two (8.3.6) and (8.3.7) allows us to derive a Reynolds lubrication type expression of the shape of the interface in terms of the external pressure and shear forces only:

$$\frac{\partial f}{\partial t} = \left(\frac{\nu_W}{3\nu_A}\right)\left[f^3 p_x\right]_x - \left(\frac{\nu_A}{2\nu_W}\right)\left[f^2\sigma\right]_x, \tag{8.3.8}$$

where subscript $x$ denotes differentiation with respect to $x$.

This result is very neat. From a model of steady flow within the droplet we have achieved a description of the nonlinear evolution of the interface, where the properties of the flow in water do not feature explicitly. No assumptions on the shape of the interface have been made and (8.3.8) applies for the later temporal stage, as the droplet becomes severely distorted. Indeed, by combining (8.3.8) with the nonlinear solution for flow in the previous section, we may bypass the flow in water altogether and obtain a two way interacting description of droplet deformation.

## 8.4 The interacting system

The algorithm for the two way interacting model of droplet deformation follows exactly that of Section 8.2, except that in this case, the shear along the interface between the fluids (8.3.4) and pressure gradient are determined explicitly the end of each time step. These results, $p_x$ and $\sigma$ are used in the finite difference version of (8.3.8) to yield the new interface shape:

$$f^{\text{new}} = \delta t \left( \left( \frac{\nu_W}{3\nu_A} \right) [f^3 p_x]_x - \left( \frac{\nu_A}{2\nu_W} \right) [f^2 \sigma]_x \right) + f. \tag{8.4.1}$$

To test the algorithm we first apply it to the Witch of Agnesi hump previously studied in Section 8.2 and Chapter 6. Results may be seen in figure (8.3), where the numerics were calculated using $\nu_A/\nu_W = 1$, $\delta x = 10^{-2}$, $\delta t = 10^{-4}$ and 128 Fourier coefficients. It is worth noting that the results are displayed in triple deck coordinates, in reality the droplet would be stretched in the $x$ direction compared to what is shown here. We see from the figure that the droplet begins to move into the area of low pressure shown in figure (8.2a) and is offset to the right by the shear forces shown in (8.2b). In figure (8.3b) we also include a graph of the values of $f(-0.5)$ and $f(0.5)$ against $t$, clearly demonstrating that, as expected, the shape of the droplet changes linearly with time.

We notice however in figure (8.3a) that the simulation appears to hold only for a short time

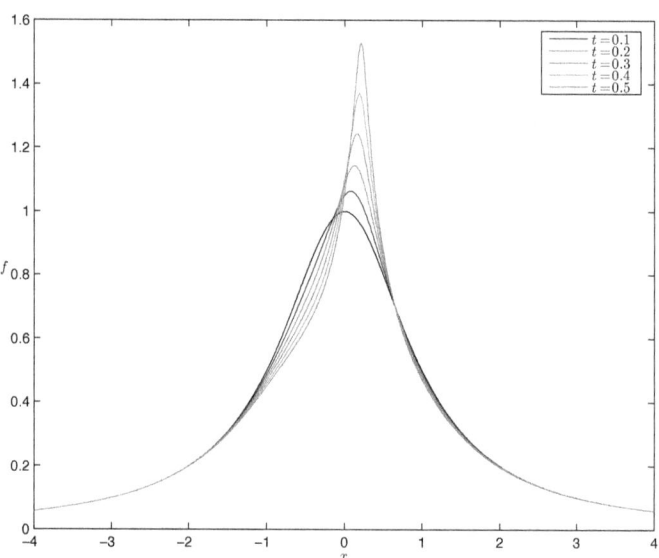

(a) Deformation of a Witch of Agnesi droplet.

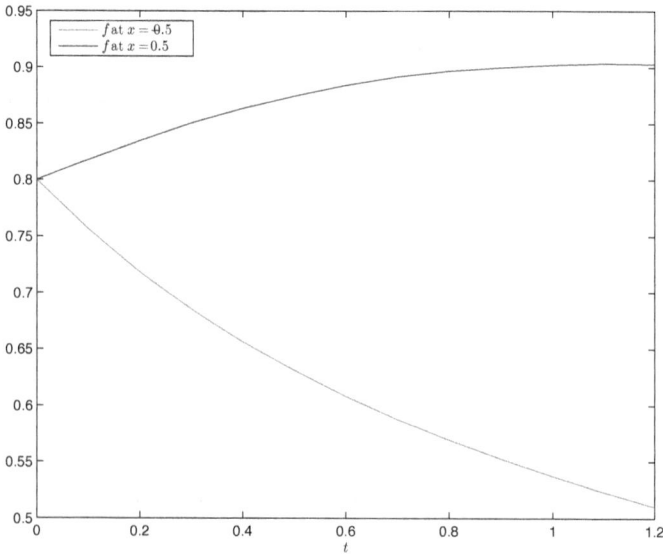

(b) Points along the interface $f(x = -0.5)$ and $f(x = 0.5)$ against time, clearly demonstrating initial linear growth of the interface.

Figure 8.3: Time dependent deformation results for a Witch of Agnesi droplet.

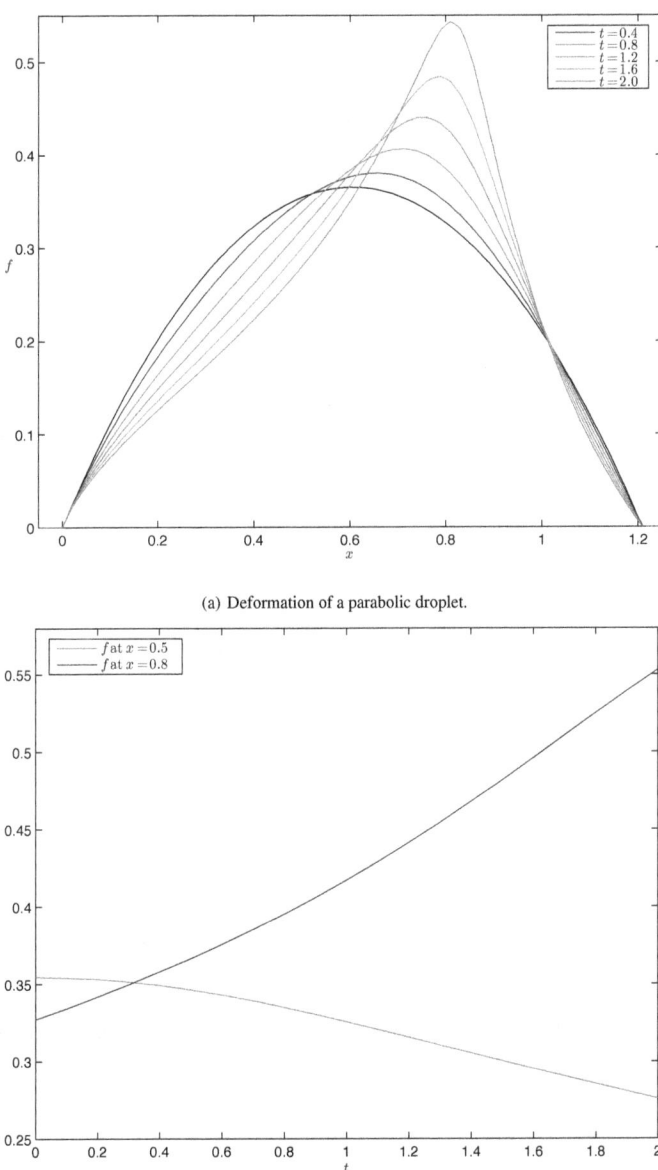

(a) Deformation of a parabolic droplet.

(b) Points along the interface $f(x = 0.5)$ and $f(x = 0.8)$ against time.

Figure 8.4: Time dependent deformation results for a parabolic droplet.

(although, of course, the time scale is generally shorter in the triple deck scaling) before

a peak appears at the top of the droplet and the solution blows up. Although not shown, this

peak continues to grow until around $t = 1.2$, after which no solution may be found numerically.

To test the algorithm further, we also include in figure (8.4a) an example of the method applied

to the parabolic hump seen in Section 8.2 and Chapter 6. In this case 512 Fourier coefficients

were used, although all other constants remain as before. Again, the droplet begins to move

into the area of low pressure shown in (8.1a), as one would expect. Figure (8.4b), a plot of the

interface at $x = 0.5$ against $t$ also demonstrates the linear growth of $f$ in time. After $t = 2$

however, this solution also breaks down, showing a similar peak as that in the Witch of Agnesi

example.

It could be that this breakdown of solution is occurring because too few Fourier coefficients are

considered. The parabolic hump has sharp corners and with $512$ coefficients, a small amount

of sinusoidal disturbance was demonstrated upstream of the droplet in the wall shear in figure

(8.1b). To confirm that the Fourier coefficients are not the cause of the blowup, we include a

final test case: a droplet with initial shape

$$f = \begin{cases} \dfrac{(1 - x^2)^4}{10} & |x| < 1, \\ 0 & \text{otherwise,} \end{cases} \qquad (8.4.2)$$

where $512$ Fourier coefficients are used. This example has a gently increasing droplet height

so that reliable results should be possible for lower $N_k$ than considered. The results for the

shape deformation are shown in figure (8.5) and a plot of the interface shape at $f = 0$, $f = 0.2$

against time is included in figure (8.5). We also show the time dependent results for shear along

the interface, figure (8.6a), and pressure figure (8.6).

The simulations for this case appear to hold for a longer time scale than the previous two

examples - up to around $t = 5$ as opposed to $t = 1.2$ or $t = 2$. Once again, we notice the linear relationship of movement of the interface with time across the majority of the period considered. However, between $t = 4$ and $t = 5$, the shear along the interface between the two fluids begins to distort causing a necking in the droplet shape. Running the algorithm to larger times causes the droplet to form a similar peak to that in both the parabolic and the Witch of Agnesi droplets.

In all the examples studied, the later stage model appears to break down at a finite time. We suspect this break-down may caused by a nonlinear localised singularity, possibly a singular pressure gradient or pressure discontinuity such as those described in Smith [56]. If this is the case, a shorter time scale would be required for further analysis. This shorter time scale would have $t = T/\epsilon + O(q)$ say, where $q << 1/\epsilon$, $T = O(1)$. Unfortunately, due to the time constraints of the project, analysis of this kind was not attempted for the current problem and is left as further work which we hope to attempt in future.

Despite this, we believe that the main point still stands overall: that the small ratios approximation may be applied within a triple deck structure to model the non linear deformation of a liquid droplet or film.

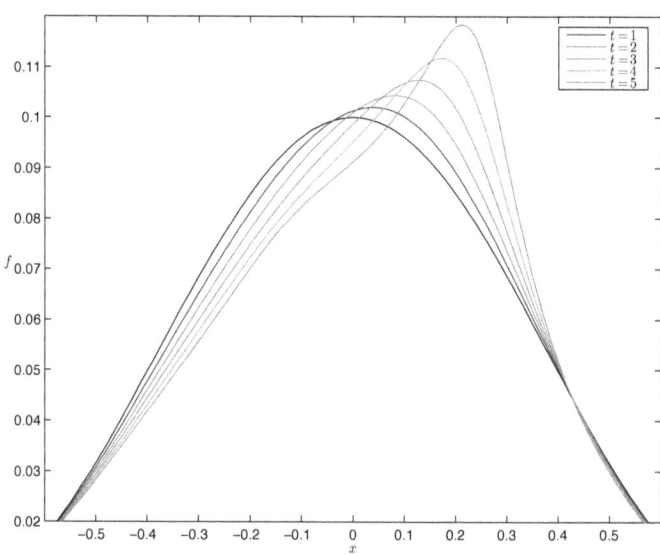

(a) Deformation of a droplet with shape described in (8.4.2).

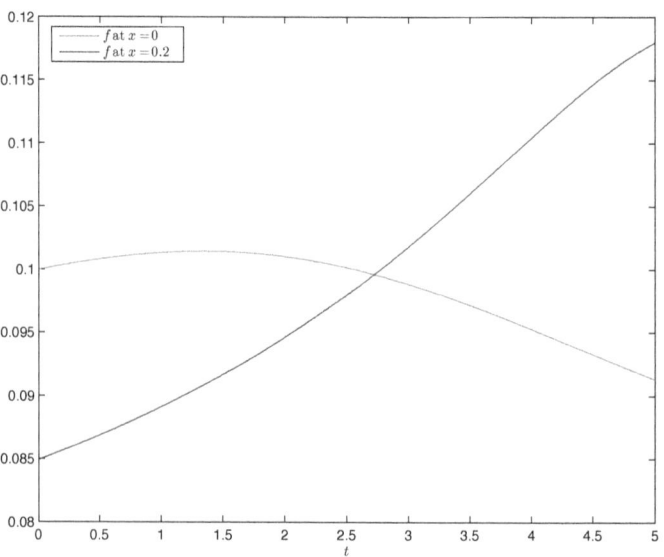

(b) Points along the interface $f(x = 0)$ and $f(x = 0.2)$ against time.

Figure 8.5: Results for the droplet with shape described in (8.4.2)

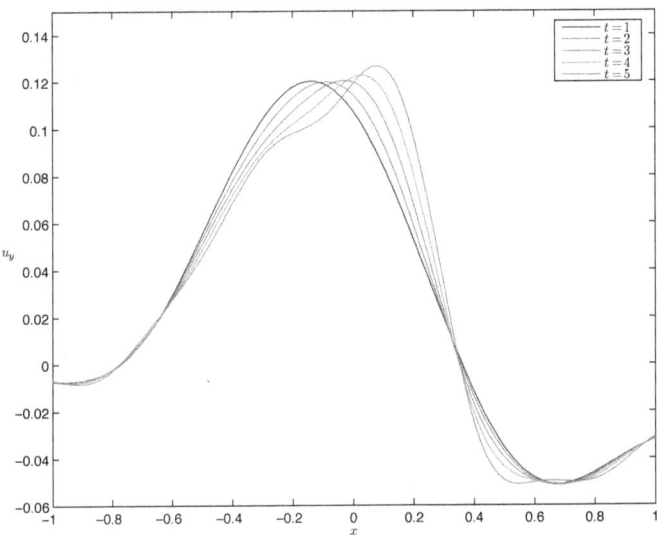

(a) Shear along the interface between the two fluids against $x$.

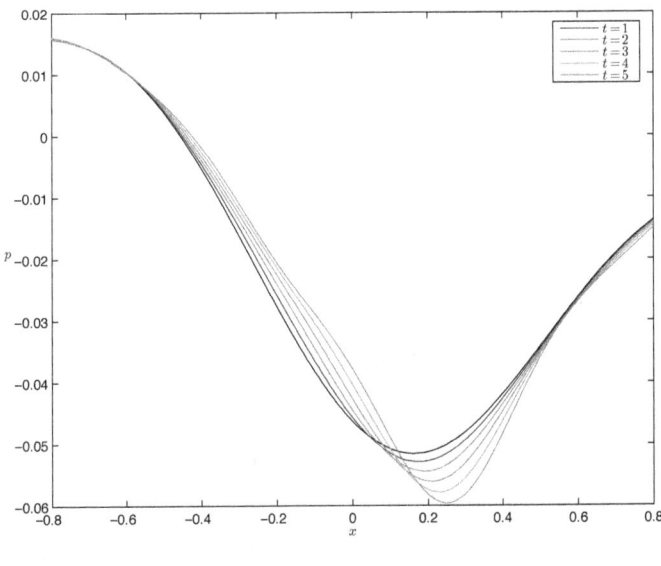

(b) Pressure against $x$.

Figure 8.6: Shear and pressure results for the droplet with shape described in (8.4.2)

Chapter 9

# Conclusions and further work

The work of this thesis has mainly been concerned with modelling the time-dependent distortion of an attached liquid droplet in the presence of an external flow field.

As described in Chapter 1, with an assumption of a small ratio between the densities of the two fluids (on which much of this work is based) the leading order velocity term in the less dense fluid will be zero along the boundary allowing us to treat the interface as solid at any given time step.

The first part of the thesis explored the application of this assumption to a wall mounted semicircular droplet in a surrounding air flow of $\mathrm{Re} = O(1)$. The model focused on the early temporal stage - while the droplet remains semicircular to leading order - allowing us to treat the flow in air as an isolated problem.

Using the lid driven cavity as a test for our numerical procedure, in Chapter 2 we derived an algorithm to numerically determine the time-dependent solution for flow in air over a surface mounted semicircular obstacle. The results showed the length of the downstream eddy increasing with Reynolds number and provided a matrix of time-dependent pressure and vorticity values along the curved wall of the semicircle for use in the interacting problem of Chapter 4.

The focus of Chapter 3 was the Stokes or creeping flow within the water droplet subject to idealised boundary conditions along the free surface. To better understand the dynamics, two model problems were considered. In the first, the circular seed, the velocities within the droplet became steady with time and the kinematic condition was found to become nonlinear at $t = O(\epsilon^{-1})$. The square droplet was analysed second and, while limited in its application to a real world problem, served well as a Cartesian version of the semicircular droplet. A series solution, steady state simulation and time dependent algorithm for the square droplet all pointed towards linear velocity growth with time and a later temporal stage which came into effect at $t = O(\epsilon^{-1/2})$. This result was mirrored in the semicircular droplet case examined towards the end of Chapter 3.

In Chapter 4 the results of the previous two chapters were combined to achieve a one-way interacting model of droplet deformation. In this example the velocities also demonstrated a linear growth with time causing, via the kinematic condition, an interface which grew like $t^2$. It was proposed that this result was due to the choice to take continuous pressure and vorticity as interfacial conditions and that applying the full stress conditions would instead see these large velocities saturate over time. Aside from including the full interface conditions, the method itself appears to be promising. An algorithm which solved for air flow over any interface shape would allow the model to simulate the later temporal stage and, perhaps, the onset of sliding rolling and droplet break-up.

A side investigation into Stokes flow near a wall, theoretically relevant to the semicircular droplet deformation model of Chapters 2 - 4 was the focus of Chapter 5, where the behaviour of the vorticity near a contact point of air, water and solid was explored. In the context of the model studied the vorticity was seen to act like $-(1 - x)^{-1/2}$ near $x = 1$ suggesting a small region of rapid clockwise rotation close to the point of contact. Extending this result to a more general form and studying the effect which the large negative vorticity has on the rest of the droplet would be an interesting problem for future work.

In the second part of the thesis we switched our attention to a droplet contained within the boundary layer of an external air flow. Chapter 6 outlines an algorithm to numerically determine the linear flow over any shaped obstacle or wall disturbance, while Chapter 8 extends this result into a nonlinear solution.

The work of Chapter 6 presented an opportunity to derive an expression for the interface between the two fluids, based on a lubrication approximation rather than the small ratios assumption. Deriving and exploring this expression was the focus of Chapter 7 where the effects of gravity, surface tension and shear forces were singled out and examined. For a droplet with pinned contact points and fixed volume, a combination of analytical an numerical techniques demonstrated the influence of the surface tension coefficient in pulling the droplet into a cylindrical shape and minimising the length of the free surface (in a two-dimensional sense). We also showed that gravity acts to flatten the droplet and that shear forces cause the droplet to deform into an asymmetrical shape offset in the same direction as the external flow field. A droplet could not satisfy the boundary conditions without the presence of surface tension (at least in end layers close to the contact points) and a restriction on the minimum gravitational force for which a droplet can withstand a shear force was found. In the case of no surface tension within the core, this was shown to be $\omega \geq 3L/2H_0^2$ for a droplet of length $L$ and height $H_0$ at $x = L/2$. The exact nature of this restriction for the full system however appears to be hidden within the nonlinearity of the system. This method too seems promising, in particular its ability to model liquid films. An extension to include time dependency would be the natural next step for this problem.

Replacing the obstacle in Chapter 6 with a second fluid and applying the small ratios assumption allowed us to achieve a two-way interacting model of droplet deformation in Chapter 8 which holds for the later temporal stage. This model takes the full stress conditions (to leading order) into account and demonstrates an interface shape which grows like $t$, as predicted in Chapter 4. More work is required to understand the cause of the instabilities which the results suggested, however, the generalised nature of the algorithm, in that it can handle liquid films or droplets of any initial shape within a boundary layer, is an exciting prospect with much potential. Building in the effects of surface tension and gravity to the model would appear to be possible, as would including surface roughness, humps or wells along the wall to which the droplet is attached. These additional features would make the model directly applicable to both aircraft icing and the treatment of aneurysms, as described in Chapter 1, along with many other industrial and medical applications.

# References

[1] M. J. Ablowitz and A. S. Fokas. *Complex variables: introduction and applications*. Cambridge Texts in Applied Mathematics. Cambridge University Press, 2003.

[2] M. Abramowitz and I. A. Stegun. *Handbook of mathematical functions with formulas, graphs, and mathematical tables*, volume 55 of *National Bureau of Standards Applied Mathematics Series*. 1964.

[3] D. M. Anderson, G. B. McFadden, and A. A. Wheeler. Diffuse-interface methods in fluid mechanics. *Annual Review of Fluid Mechanics*, 30:pp. 139–165, 1998.

[4] G. K. Batchelor. *An introduction to fluid dynamics*. Cambridge University Press, 1967.

[5] B. J. Bentley and L. G. Leal. An experimental investigation of drop deformation and breakup in steady, two-dimensional linear flows. *Journal of Fluid Mechanics*, 167:pp. 241–283, 1986.

[6] S. Bhattacharyya, S. C. R. Dennis, and F. T. Smith. Separating shear flow past a surface-mounted blunt obstacle. *Journal of Engineering Mathematics*, 39:pp. 47–62, 2001.

[7] F. Bierbrauer and T. N Phillips. The numerical prediction of droplet deformation and break-up using the godunov marker-particle projection scheme. *Int. J. Numer. Meth. Fluids*, 56:pp. 1155–1160, 2008.

[8] H. Blasius. Grenzschichten in flussigkeiten mit kleiner reibung. *Journal of Mathematical Physics*, 56:pp. 1–37, 1908.

[9] O. R. Burggraf. Analytical and numerical studies of the structure of steady separated flows. *Journal of Fluid Mechanics*, 24:pp. 113–151, 1966.

[10] G. F. Carrier, M. Krook, and C. E. Pearson. *Functions of a complex variable*. Society for Industrial and Applied Mathematics (SIAM), Philadelphia, PA, 2005.

[11] H. S. Carslaw. *Introduction to the Theory of Fourier's Series and Integrals, 3rd ed.* Dover, New York, 1950.

[12] R. Croce, M. Griebel, and M. A. Schweitzer. Numerical simulation of bubble and droplet deformation by a level set approach with surface tension in three dimensions. *International Journal for Numerical Methods in Fluids*, 62:pp. 963–993, 2010.

[13] S. C. R. Dennis and F. T. Smith. Steady flow through a channel with a symmetrical constriction in the form of a step. *Proceedings of the Royal Society. London. Series A. Mathematical and Physical Sciences*, 372(1750):pp. 393–414, 1980.

[14] P. Dimitrakopoulos. Deformation of a droplet adhering to a solid surface in shear flow: onset of interfacial sliding. *Journal of Fluid Mechanics*, 580:pp. 451–466, 2007.

[15] E. B. Dussan V. On the ability of drops to stick to surfaces of solids. part 3. the influences of the motion of the surrounding fluid on dislodging drops. *Journal of Fluid Mechanics*, 174:pp. 381–397, 1987.

[16] E. B. Dussan V. and S. H. Davis. On the motion of a fluid-fluid interface along a solid surface. *Journal of Fluid Mechanics*, 65:pp. 71–95, 1974.

[17] L. Fermo and M. G. Russo. Numerical methods for Fredholm integral equations with singular right-hand sides. *Advances in Computational Mathematics*, 33:pp. 305–330, 2010.

[18] U. Ghia, K. N. Ghia, and C. T. Shin. High-Re solutions for incompressible flow using the Navier-Stokes equations and a multigrid method. *Journal of Computational Physics*, 48:pp. 387–411, 1982.

[19] S. Goldstein. Concerning some solutions of the boundary layer equations in hydrodynamics. *Mathematical Proceedings of the Cambridge Philosophical Society*, 26:pp. 1–30, 1930.

[20] A.K. Gupta and S. Basu. Deformation of an oil droplet on a solid substrate in simple shear flow. *Chemical Engineering Science*, 63:pp. 5496 – 5502, 2008.

[21] F. H. Harlow and J. E. Welch. Numerical calculation of time-dependent viscous incompressible flow of fluid with free surface. *Physics of Fluids*, 8:pp. 2182–2189, 1965.

[22] P. D. Hicks and R. Purvis. Air cushioning and bubble entrapment in three-dimensional droplet impacts. *Journal of Fluid Mechanics*, 649:pp. 135–163, 2010.

[23] J. J. L. Higdon. Stokes flow in arbitrary two-dimensional domains: shear flow over ridges and cavities. *Journal of Fluid Mechanics*, 159:pp. 195–226, 1985.

[24] C. W. Hirt and B. D. Nichols. Volume of fluid (vof) method for the dynamics of free boundaries. *Journal of Computational Physics*, 39:pp. 201–225, 1981.

[25] M. Ishii. Thermo-fluid dynamic theory of two-phase flow. *NASA STI/Recon Technical Report A*, 75, 1975.

[26] C. E. Jobe and O. R. Burggraf. The Numerical Solution of the Asymptotic Equations of Trailing Edge Flow. *Proceedings of the Royal Society of London. A. Mathematical and Physical Sciences*, 340:pp. 91–111, 1974.

[27] D. D. Joseph and Y. Y. Renardy. *Fundamentals of two-fluid dynamics. Part I*, volume 3. Springer-Verlag, New York, 1993. Mathematical theory and applications.

[28] D. D. Joseph and Y. Y. Renardy. *Fundamentals of two-fluid dynamics. Part II*, volume 4. Springer-Verlag, New York, 1993. Lubricated transport, drops and miscible liquids.

[29] H. B. Keller and H. Takami. Numerical studies of steady viscous flow about cylinders. In *Numerical Solutions of Nonlinear Differential Equations*, pages pp. 115–140. John Wiley & Sons Inc., 1966.

[30] W. E. Langlois. *Slow viscous flow*. The Macmillan Co., New York, 1964.

[31] X. Li and C. Pozrikidis. Shear flow over a liquid drop adhering to a solid surface. *Journal of Fluid Mechanics*, 307:pp. 167–190, 1996.

[32] M. J. Lighthill. On boundary layers and upstream influence. ii. supersonic flows without separation. *Proceedings of the Royal Society of London. Series A, Mathematical and Physical Sciences*, 217:pp. 478–507, 1953.

[33] D. B. Macleod. On a relation between surface tension and density. *Transactions of the Faraday Society*, 19:38–41, 1923.

[34] V. V. Melesko and A. M. Gomilko. Two-dimensional Stokes flow in a semicircle. *International Journal of Fluid Mechanics Research*, 27:pp. 56–61, 2000.

[35] A. F. Messiter. Boundary-layer flow near the trailing edge of a flat plate. *SIAM Journal on Applied Mathematics*, 18:pp. 241–257, 1970.

[36] S. G. Mikhlin. *Integral equations and their applications to certain problems in mechanics, mathematical physics and technology*. Pergamon Press, New York, 1957.

[37] H. K. Moffatt. Viscous and resistive eddies near a sharp corner. *Journal of Fluid Mechanics*, 18:pp. 1–18, 1964.

[38] N. I. Muskhelishvili. *Singular integral equations*. Dover Publications Inc., New York, 1992.

[39] T. G. Myers. Thin films with high surface tension. *SIAM Review*, 40:pp. 441–462, 1998.

[40] T. G. Myers, J. P. F. Charpin, and C. P. Thompson. Slowly accreting ice due to supercooled water impacting on a cold surface. *Physics of Fluids*, 14:pp. 240–256, 2002.

[41] S. Osher and R. Fedkiw. *Level set methods and dynamic implicit surfaces*, volume 153 of *Applied Mathematical Sciences*. Springer-Verlag, New York, 2003.

[42] M. K. Politovich. Aircraft Icing Caused by Large Supercooled Droplets. *Journal of Applied Meteorology*, 28:pp. 856–868, 1989.

[43] C. Pozrikidis. *Boundary integral and singularity methods for linearized viscous flow*. Cambridge University Press, 1992.

[44] L. Prandtl. On motion of fluids with very little viscosity. In *Third International Congress of Mathematics*, 1904.

[45] H. A. Priestley. *Introduction to complex analysis*. Oxford University Press, 2003.

[46] M. C. Pugh. Notes on blowup and long wave unstable thin film equations. *Mathematical Sciences research institute*, pages pp. 65–74, 2006.

[47] Lord J. W. S. Rayleigh. *Scientific papers*. Six volumes bound as three. Vol. 6:1911-1919. Dover Publications Inc., New York, 1964.

[48] M. Renardy, Y. Renardy, and J. Li. Numerical simulation of moving contact line problems using a volume-of-fluid method. *Journal of Computational Physics*, 171:pp. 243–263, 2001.

[49] Y. Y. Renardy, M. Renardy, and V. Cristini. A new volume-of-fluid formulation for surfactants and simulations of drop deformation under shear at a low viscosity ratio. *European Journal of Mechanics - B/Fluids*, 21:pp. 49–59, 2002.

[50] D. P. Rizzetta, O. R. Burggraf, and R. Jenson. Triple-deck solutions for viscous supersonic and hypersonic flow past corners. *Journal of Fluid Mechanics*, 89:pp. 535–552, 1978.

[51] L. Rosenhead. *Laminar boundary layers*. Oxford : Clarendon Press, 1963.

[52] A.D Schliezer and R.T. Bonnecaze. Displacement of a two-dimensional immiscible droplet adhering to a wall in shear and pressure-driven flows. *Journal of Fluid Mechanics*, 383:pp. 29–54, 1999.

[53] D. Sivakumar and C. Tropea. Splashing impact of a spray onto a liquid film. *Physics of Fluids*, 14:pp. 85–88, 2002.

[54] F. T. Smith. Laminar flow over a small hump on a flat plate. *Journal of Fluid Mechanics*, 57:pp. 803–824, 1973.

[55] F. T. Smith. Upstream interactions in channel flows. *Journal of Fluid Mechanics*, 79:pp. 631–655, 1977.

[56] F. T. Smith. Finite-time break-up can occur in any unsteady interacting boundary layer. *Mathematika. A Journal of Pure and Applied Mathematics*, 35:pp. 256–273, 1988.

[57] F. T. Smith, P. W. M. Brighton, P. S. Jackson, and J. C. R. Hunt. On boundary-layer flow past two-dimensional obstacles. *Journal of Fluid Mechanics*, 113:pp. 123–152, 1981.

[58] F. T. Smith and O. R. Burggraf. On the development of large-sized short-scaled disturbances in boundary layers. *Proceedings of the Royal Society of London. Series A, Mathematical and Physical Sciences*, 399:pp. 25–55, 1985.

[59] F. T. Smith, L. Li, and G. X. Wu. Air cushioning with a lubrication/inviscid balance. *Journal of Fluid Mechanics*, 482:pp. 291–318, 2003.

[60] F. T. Smith and J. H. Merkin. Triple-deck solutions for subsonic flow past hymps, steps, concave or convex corners and wedged trailing edges. *Computers & Fluids*, 10:pp. 7–25, 1982.

[61] F. T. Smith, N. Ovenden, and R. Purvis. Industrial and biomedical applications. In *IUTAM Symposium on One Hundred Years of Boundary Layer Research*, volume 129 of *Solid Mechanics and Its Applications*, pages pp. 291–300. Springer Netherlands, 2006.

[62] F. T. Smith and R. Purvis. Air effects on droplet impact. *4th AIAA Theoretical Fluid Mechanics Meeting, Toronto, Canada*, 2005.

[63] I. J. Sobey. *Introduction to interactive boundary layer theory*, volume 3 of *Oxford Texts in Applied and Engineering Mathematics*. Oxford University Press, Oxford, 2000.

[64] P. Spelt. Shear flow past two-dimensional droplets pinned or moving on an adhering channel wall at moderate reynolds numbers: a numerical study. *Journal of Fluid Mechanics*, 561:pp. 439–463, 2006.

[65] K. Stewartson. On the impulsive motion of a flat plate in a viscous fluid. *The Quarterly Journal of Mechanics and Applied Mathematics*, 4:pp. 182–198, 1951.

[66] K. Stewartson. On the flow near the trailing edge of a flat plate. *Proceedings of the Royal Society of London. Series A, Mathematical and Physical Sciences*, 306:pp. 275–290, 1968.

[67] K. Stewartson. On laminar boundary layers near corners. *The Quarterly Journal of Mechanics and Applied Mathematics*, 23:pp. 137–152, 1970.

[68] M. Sussman, A. S. Almgren, J. B. Bell, P. Colella, L. H. Howell, and M. L. Welcome. An adaptive level set approach for incompressible two-phase flows. *Journal of Computational Physics*, 148:pp. 81–124, 1999.

[69] V. V. Sychev, A. I. Ruban, V. V. Sychev, and G. L. Korolev. *Asymptotic theory of separated flows*. Cambridge University Press, Cambridge, 1998.

[70] D. P. Telionis. *Unsteady viscous flows*. Springer-Verlag, New York, 1981.

[71] S. K. Thomas, R. P Cassoni, and C. D MacArthur. Aircraft anti-icing and de-icing techniques and modeling . *Journal of Aircraft*, 33:pp. 841–854, September 1996.

[72] E. Villermaux and B. Bossa. Single-drop fragmentation determines size distribution of raindrops. *Nat Phys*, 5:pp. 697–2473, 2009.

[73] A. White. Wall-shape effects on multiphase flow in channels. *Theoetical and Computational fluid dynamics*, 2010.

[74] F. M. White and I. Corfield. Viscous fluid flow. 1974.

[75] L. C. Woods. A note on the numerical solution of fourth order differential equations. *Aerodynamics Quarterly*, 5:pp. 176–184, 1954.